The Institute of Biology's
Studies in Biology no. 17

Fungal Parasitism

by *Brian J. Deverall* Ph.D.

Lecturer, Botany Department, Imperial College of
Science and Technology, University of London

New York   CRANE, RUSSAK & COMPANY, INC.
347 Madison Avenue
New York, New York 10017

© Brian J. Deverall, 1969

First published 1969

First published in
the United States of America in 1969

First published in Great Britain by
Edward Arnold (Publishers) Ltd.

Library of Congress Catalog Card Number: 73-83206

Printed in Great Britain by
William Clowes and Sons Ltd, London and Beccles

General Preface to the Series

It is no longer possible for one textbook to cover the whole field of Biology and to remain sufficiently up to date. At the same time students at school, and indeed those in their first year at universities must be contemporary in their biological outlook and know where the most important developments are taking place.

The Biological Education Committee, set up jointly by the Royal Society and the Institute of Biology, is sponsoring, therefore, the production of a series of booklets dealing with limited biological topics in which recent progress has been most rapid and important.

A feature of the series is that the booklets indicate as clearly as possible the methods that have been employed in elucidating the problems with which they deal. There are suggestions for practical work for the student which should form a sound scientific basis for his understanding.

1969   INSTITUTE OF BIOLOGY
                                                                         41 Queen's Gate
                                                                         London, S.W.7

Preface

Fungal parasitism is a mode of life which is of interest to all students of biology. The mechanism of evolution of the intricate relationships between structures and life cycles of parasites and hosts are major challenges to the imagination. Two of the most interesting areas of biological research concern the nature of obligate parasitism of fungi on plants and the great degree of specificity which often exists between parasites and hosts.

This booklet is especially concerned with fungal parasitism on plants. Because of the economic importance of plant diseases caused by fungi, a great deal of information is available about the habit. Not only has the relative ease of experimentation with plant disease provided much of this information but it also makes classroom experiments simpler than would be the case with parasitism on animals. However, experiments in school on fungal parasitism are several steps more difficult than work with single organisms. Therefore worthwhile observations in the garden and field are suggested, to be integrated with experimental studies.

London   B.J.D.
1969

Contents

Ecology and Nutrition of Fungi 1

1.1 Parasitism

Parasitism is the mode of life of an organism which lives on or in another living organism. The parasite derives at least part of its food from the host, and perhaps shelter. Most living organisms are parasitized at some stage of their life cycle by other organisms. Most parasites are fungi, bacteria, protozoa, flat worms, round worms and insects, but many groups of organisms contain a few parasitic species. The cuckoo is a parasitic bird as a nestling when it derives its food from its unwitting foster-parents. Lampreys are parasitic fish which attach to, and suck blood from other fish. A few higher plants are parasitic, such as the achlorophyllous dodder and broomrape which obtain all their organic food from their host plants and the chlorophyllous mistletoe which takes salts and water from its host plant.

1.2 Modes of life of fungi

Fungi have two major ways of obtaining their food, by saprophytism and parasitism. Saprophytic fungi either colonize decaying plant and animal material or absorb soluble organic substances which have been released from living or dead organisms. Parasitic fungi grow on or in living plants and animals, and often damage their hosts. When damage occurs, the host is said to be diseased and the parasite is termed a pathogen. Lichens and mycorrhizal roots are products of symbiotic relationships between fungi and plants, in which both partners derive benefit from the association. Symbiosis is an extreme form of the balanced relationship which permits a parasitic fungus to live in a degree of harmony with its host until the fungus has grown and reproduced itself.

1.3 Nutritional needs of fungi

It should be clear now that fungi resemble animals and differ from green plants in that they need to be supplied with food in an already elaborated organic form. Their nutrition resembles that part of animal nutrition which involves the absorption of soluble substances.

Most saprophytic fungi and a large number of parasitic fungi can be grown in pure culture on artificial media. The medium must provide a soluble utilizable carbohydrate, a source of nitrogen and certain salts plus water at a pH near 6·0. Glucose satisfies the carbohydrate needs of all fungi, although some fungi can utilize several or all of the other naturally occurring hexose and pentose sugars and some can grow if supplied with polymers such as starch or cellulose. All fungi will accept a protein hydro-lyzate as a source of nitrogen, most will grow well if l-asparagine is the only source, many can use ammonium ions and some can grow with nitrate ions as the only nitrogen source. Inorganic requirements are for the following ions: magnesium, potassium, phosphate, sulphate and, in trace amounts, iron, zinc, copper, manganese and molybdenum. Many fungi grow and produce reproductive bodies and spores in the dark at 20–25°C on simple media of this type. Some fungi require to be supplied with traces of certain vitamins for growth and/or reproduction, the most common needs being for biotin and thiamine (vitamins H and B_1).

With some minor exceptions, it must be emphasized that saprophytes and many parasites demand no more than these simple nutrients from their substrates and hosts, respectively. Parasitic fungi which can be grown in artificial culture are termed facultative parasites, implying that although they are ecological parasites they are nutritional saprophytes. Their parasitic habit is concerned with the exploitation of a specialized ecological niche which aids them in competition for simple nutrients. Their special features result from the evolution of structures and repro-ductive cycles which permit them to invade, derive nutrients from and time their seasonal changes with their living hosts. As will become apparent in the course of this booklet, parasitic fungi have to be able, in a little understood way, to overcome mechanisms in the host which may resist their development. Obligate parasites are fungi which need to be associated with the living cells of their specific host in order to grow and reproduce. They cannot be cultured on artificial media, for reasons as yet unknown.

Grades of parasitism exist within the categories of facultative and obligate parasite. Some parasites may grow in nature both parasitically on senescing but living tissues and saprophytically on dead matter. Other facultative parasites grow and reproduce freely on their hosts, but grow slowly and reproduce poorly or not at all on artificial media. Some obligate parasites produce atypical growth on artificial media, some undergo spore germination but many fail to show any development.

Isolation, Culture and Germination of Parasitic Fungi 2

This is a chapter of methods, which can be carried out in simply equipped laboratories and which underlie most of the experiments described in later chapters.

2.1 Making media

Two essential pieces of equipment needed for making sterile artificial media for the culture of fungi are a small oven and a pressure-cooker or autoclave. The oven is used to sterilize dry glassware, such as petri dishes, by incubating them at 160°C for 2 h. The pressure-cooker is used without pressure to melt suspensions of agar in water and, after combining components with the melted agar, to sterilize the medium at a pressure of 15 lb/in² for 20 minutes.

Many parasitic fungi, apart from obligate parasites, will grow on ready-mixed dextrose peptone agar (Oxoid), which requires dissolving and sterilizing only. A natural extract medium which supports good growth of more fastidious facultative parasites is potato-dextrose agar, made as follows: Boil 200 g sliced, peeled potato tubers in water until soft, crush and filter through cheese cloth to give a volume of 500 ml, to which 20 g of dextrose (glucose) is added. Steam and melt 20 g agar in 500 ml water. Mix the nutrients and the agar, dispense in desired volumes into test tubes or bottles and sterilize by pressure-cooking. After sterilization, the medium may be poured into petri dishes when it has cooled to a temperature comfortable to the hand. The procedures of flaming necks of tubes and pouring, inoculating and incubating dishes of medium are illustrated in Fig. 2–1.

2.2 Obtaining cultures of parasitic fungi

Pure cultures of some facultative parasites can be obtained from the Commonwealth Mycological Institute, Ferry Lane, Kew, Surrey (see *Class Work with Fungi* by DADE and GUNNELL). The cultures are stored either freeze-dried or on agar slopes under liquid paraffin in tubes, and are started into growth by plating out as in Fig. 2–1. North American readers should see the *Sourcebook of Laboratory Exercises in Plant Pathology*, listed in Further Reading, for advice on obtaining cultures.

Some parasitic fungi can be isolated easily from diseased material by picking off pieces of aerial mycelium or spore mats with a sterile needle

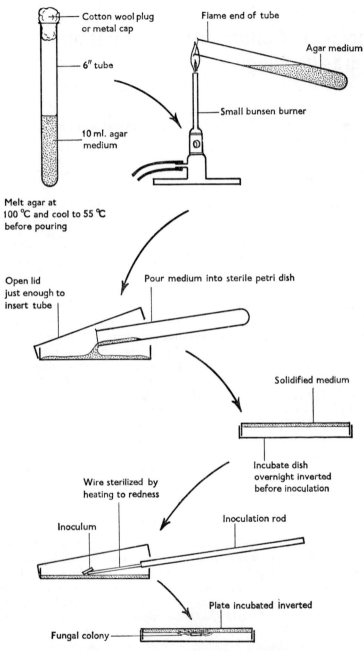

Fig. 2-1 Technique of culturing fungi on solid media in petri dishes.

or forceps and plating out. This technique is readily applied to some of the parasitic fungi which rot apple or citrus fruit, and to spores of *Botrytis* spp. on infected flower heads of several garden plants in wet weather.

A problem associated with this method, and particularly with methods directed at inducing the parasitic fungus to grow out from pieces of host tissue into media, is the avoidance of saprophytic fungi and bacteria. Saprophytic fungi may be present as resting spores on host tissue, or as secondary invaders of dying parasitized tissue. Superficial saprophytes can be killed by immersion of tissue pieces for 10 sec in 0.1% $HgCl_2$ followed by repeated rinses in changes of sterile water. Depending upon size and condition of tissue, the period and concentration of the dip may be altered to advantage. Pieces of treated tissue should then be cut up and plated out on media.

Even after using a highly selective method to isolate a fungus, several organisms may grow out on the medium. This presents the problem of separating them and deciding which one is the parasite. A way of encouraging the growth of fungi and not that of bacteria is to use acid media, prepared by adding 0.5 g malic or tartaric acid to 100 ml of melted agar. The only fungi to be favoured by alkaline conditions are some of the lower fungi. A further refinement is the addition of antibiotics, such as aureomycin, streptomycin or chloramphenicol, to the medium at a concentration of 25 μg/ml to prevent bacterial growth. The separation of several fungi is a little more difficult, and it helps if the fungi are prevented from growing rapidly. Media which provide one tenth the normal concentration of nutrients are useful in achieving this. Additionally a substance such as Rose Bengal may be included in the medium at 25 μg/ml to slow the growth of all fungi and to suppress bacteria. Attempts to separate mycelia of fungi must be made as soon as possible after they have started to grow. Fungal hyphae can be examined microscopically in closed petri dishes through the undersurface of the dish, provided that this is not heavily scratched and that there are not particles in the medium. Examination at 12, 24 and 48 h after inoculation of the medium is necessary to see the origin of hyphae. A very useful device for picking out hyphae is a dummy cutting objective, which is attached to a microscope and racked down into medium to cut out a disc of agar of the same size as the field of view under low-power magnification. The disc can then be picked from the dish on a sterile needle and plated out on fresh medium.

The dummy cutting objective is also useful in picking out single spores from a suspension of widely dispersed spores in sterile water spread over agar in a petri dish. Single-spore isolations are often used by plant pathologists both to separate the spores of several isolated organisms and to start cultures of one genetic type within a species of fungus.

Success in isolating pure cultures of fungi from plants leads to the problem of finding which organism, if any, is parasitic. Plant pathologists face this problem whenever they investigate a disease with an unknown cause.

Experience has taught that the following procedures, based on Koch's postulates concerning causes of diseases, are sound practice.

  (i) Isolate organisms from the diseased tissue.
 (ii) Culture each organism separately.
(iii) Inoculate each organism into separate healthy host plants.
(iv) Observe the development of the appropriate symptoms.
 (v) Re-isolate the suspected parasite from the affected host.

2.3 Production of parasitized plants

The production of a parasitized plant under test conditions is essential in all attempts to prove the ability of the isolated fungus to grow parasitically. Furthermore it is the only means of culturing obligately parasitic fungi, and of studying physiological responses of plants to infection by fungi. This phase of the work limits the choice of diseases for easy experimentation in the laboratory or greenhouse. It is also an occasion to imagine the problems presented to the plant pathologist first confronted with, for example, an important disease of tree foliage caused by a suspected obligate parasite.

The most important conditions for success in producing diseased plants are emphasized here, and some of the simpler procedures are described below. Fungal spores germinate in very humid atmospheres or in water droplets. With very few exceptions, moist conditions must be maintained during germination and entry into a plant. Furthermore, successful development of the fungus inside the plant requires that the host tissues remain turgid. Most fungi enter plants more readily through physical wounds, and wounding is essential for many infections. Many facultative parasites can be grown on detached parts of their particular host species. Obligate parasites may not undergo their full development on excised tissues, probably because they must be supported by the metabolism of cells, which are not undergoing senescence induced by detachment.

The simplest plant diseases to produce in the laboratory are those of mature fruits or potato tubers. Rots of apples and oranges are good examples. The parasitic fungi causing these rots are wound entrants. Healthy fruit should be washed gently, and allowed to dry. Using a blade sterilized by brief flaming, a V-shaped cut 0·5 cm long should be made in the fruit surface. A piece of fungal mycelium from the advancing edge of a fungal culture should be inserted in the fruit beneath the cut tissue, which can then be pressed back into place. The fruit should be incubated in a moist chamber at room temperature for several days. A suitable moist chamber is a plastic box lined with a layer of filter paper soaked with sterile water and closed with a well-fitting lid. Most fruit rots are rapid, although internal symptoms may develop more rapidly than external ones. Sample fruit should be cut open at daily intervals to see the progress of the rot. Rotting will be followed by the sporulation of the fungus on the outer surface of the fruit. These diseases are named commonly by the appearance of the rot or of the spore mat of the parasite. Brown rot of apple caused by *Sclerotinia fructigena* and green mould of orange caused by *Penicillium digitatum* are most frequently seen, as in Plates 1 and 2.

Simple leaf diseases which can be produced in the laboratory are those of broad bean (*Vicia faba*) caused by species of *Botrytis*. A dwarf variety of broad bean should be grown in the greenhouse between 15 and 25°C to the stage when the plants bear several leaves. Plant development in winter is aided by illumination from a few fluorescent tubes suspended one foot above the pots, but artificial light is not essential. The fungus *Botrytis cinerea* or closely related species may be isolated from decaying heads of many garden flowers or senescent leaves at the base of a lettuce crop. Spores are produced on culture media after 7–10 days growth at 18–23°C, and are harvested beneath sterile water by gently scraping the spore mat with a sterile needle. The spore suspension should be filtered through layers of muslin to remove hyphal fragments. Spores may be washed free of nutrients from the culture medium by allowing them to sediment for 1 h in a test tube or by centrifuging them at 500 g for 1 min. The supernatant is discarded, and the spores are resuspended in sterile water. This procedure can be repeated once or twice, depending upon the needs of the experiment. Spore concentration should be adjusted to 10^5–10^6 spores/ml, either roughly by making the suspension just obvious to the naked eye or precisely after estimating density in an haemocytometer, Fig. 2–2. The latter is a microscope slide which bears a cell of known depth over a known marked area, and is used routinely for medical blood cell counts. Droplets of 20 µl (0·02 ml) volume of suspension should be placed with a graduated 0·1 ml pipette on the upper surface of detached leaves in a moist chamber of the type described above. Care must be taken to keep the spores in suspension during the process of pipetting. The lids of plastic boxes should be lightly sealed with a thin smear of petroleum jelly. Boxes should be incubated at 15–20°C out of direct sunlight, which may dry the leaves and droplets thus preventing infection. Brown spots should appear within 1 or 2 days beneath the infection droplets. Infection is more rapid and pronounced on leaves which have been abraded or picked with a needle.

The infectivity of *B. cinerea* can be greatly promoted by addition of orange juice to the spore suspension. The stimulant in orange juice is not known, but it enables *B. cinerea* to colonize broad bean leaves in the same way as the highly specialized bean parasite *B. fabae*. This parasite can be isolated from some brown spots on bean leaves in garden or field, and along with *B. cinerea* from spore mats on the surfaces of dead bean leaves and flowers. *B. fabae* is distinguished from *B. cinerea* by its larger spores, 20–25 µ compared with 10–15 µ in length, respectively. *B. fabae* often produces many black sclerotia, which are masses of resting hyphae, and may produce relatively few spores. Spore production is increased in daylight. Parallel infection of opposite halves of leaves with spores of *B. fabae* and *B. cinerea* in water provides a clear demonstration of the different powers of the two fungi, as in Plate 3.

Rust fungi can be grown in the greenhouse on young plants of their appropriate host species. A dust of uredospores from diseased freshly collected plants followed by a mist of water from a hand atomizer or scent spray should be allowed to settle over the leaf surfaces. The plants should

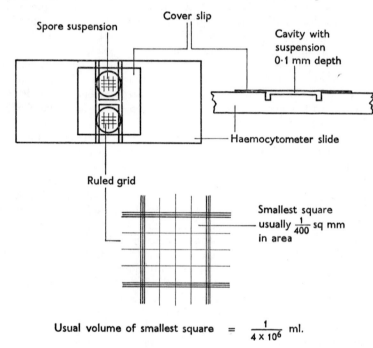

Fig. 2-2   Diagram of haemocytometer slide.

be incubated beneath a polythene hood over a tray of water at 15–18°C for 24 h and then returned to the greenhouse bench. First symptoms of rust infections are pale flecks, appearing within a week, to be followed in the second week by the eruption of rust pustules from the flecks. Common rusts which can be cultured in this way are *Antirrhinum* rust, broad bean rust and cereal rusts such as crown rust of oats, black stem rust of wheat and brown leaf rust of wheat.

Diseases of roots, stems at soil level and the vascular systems of plants can be reproduced in the greenhouse by combining a macerate of the mycelium of the parasite with the soil, sand or vermiculite in which the plant has grown. Sufficient mycelium can be produced in liquid media (i.e. media without agar) dispensed in 40 ml volumes in 12-ounce medicine bottles (medical flats), plugged with cotton wool or fitted with slightly slackened screw caps. Small pieces of fungus from agar cultures are added to the bottles, which are then incubated horizontally at room temperature for a week. The mycelium may be macerated in a pestle and mortar with sand or in a food mixer. Damping-off diseases of cress, cabbage or lettuce seedlings can be caused by species of *Pythium* isolated from soil, *Botrytis cinerea* or *Rhizoctonia solani* within a week of adding the mycelial macerate to the soil. Incubating the seedlings under a polythene hood over a tray of water facilitates parasitic attack.

2.4   Germination of spores of parasitic fungi

The spores of many fungi can be germinated on microscope slides or cover slips in water or nutrient solutions. The germination test is used in studying nutrient requirements and behaviour of spores and germ tubes, and in screening substances for fungicidal action.

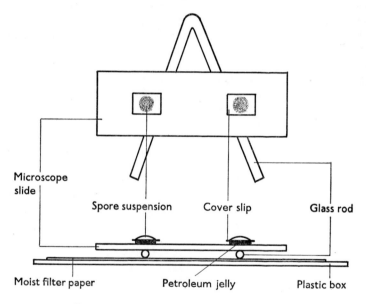

Fig. 2–3   Technique of spore germination test.

Spores are obtained from diseased plants or as stated above from *Botrytis* in culture. *Botrytis cinerea* is again an excellent organism for study, because it produces many spores, which are easily handled. The spores are very responsive to traces of glucose or sucrose, but will germinate in water alone in 24 h at concentrations less than one million per ml.

Uredospores of rust fungi do not require nutrients and germinate within 12 h in distilled water.

The main problems in germination tests concern the cleanliness of glass surfaces and the density of the spore suspension. Clean glass surfaces are provided by new cover slips, which should be lifted by clean forceps and rested horizontally on smears of petroleum jelly on microscope slides. Twenty μl (0·02 ml) drops of spore suspension are placed on upper surfaces of the cover slips, and the slides are incubated for 24 h at room temperatures on moist filter papers in petri dishes, as in Fig. 2–3. Fungal spores can affect the germination of their neighbours in dense spore suspensions. The most common effect is mutual inhibition by unknown factors, but stimulations have been noted. Wise precautions are to ensure that spores in a germination droplet are separated by several spore lengths, and to increase dilution of suspensions in which germination does not occur. Many spores are sticky or difficult to wet, and their suspension and dispersion may be aided by low concentrations (0·1 ml/100 ml) of non-toxic wetting agents, such as Tween 80.

After germination, it is helpful to add a small drop of cotton blue in lactophenol to the droplets before examining spores under low-power magnification of the microscope. The percentage of spores which have germinated and the lengths of germ-tubes, perhaps as multiples of spore lengths, may be estimated.

In addition to the stimulatory effects of glucose or plant sap on the germination of spores of *Botrytis*, the fungicidal effect of copper ions can easily be demonstrated. The mid-points in a concentration range of 10–500 μg/ml of copper sulphate will be found to suppress germination. This is the basis of the fungicidal action of Bordeaux mixture, which was the first fungicide to be formulated (in the 1880s) and which is still one of the most effective.

3.1 Taxonomic grouping

Most groups of fungi contain some parasitic species, but most of the parasitic fungi are members of a limited number of taxonomic orders. Within these orders, the component species are not only, of course, similar morphologically, but they usually cause similar types of disease. It is quite easy for a mycologist to identify a fungus as a member of the rusts by microscopic examination of the spores or by looking at diseased plants a yard away. In fact closely related species of parasitic fungi are often distinguished more easily by noting the host and any symptoms (signs) of disease than by looking at the morphology of the fungus. This situation has dangers as will become apparent later

3.2 Lower fungi

3.2.1 Slime-moulds

Slime-moulds of different types are classified with the fungi for convenience, although biologists are doubtful concerning their evolutionary affinity with fungi, algae and protozoa. Several of these organisms are parasitic on plants, and the best known is *Plasmodiophora brassicae*, which causes the widespread and troublesome disease known as club-root. Many gardeners experience difficulty in repeatedly growing cabbage and its relatives in the Cruciferae including wallflower. To check for club-root, plants which have been making poor growth should be examined for nodules or warty outgrowths on the roots, which may become extremely swollen. (See Plate 10.) Sections of these areas examined under the microscope should show that host cells have grown greatly in size, and it may be possible by staining with Feulgen's stain to find that host nuclei are up to 10 times the normal size in some of these cells. In many cells, masses of protoplasm of the parasite should be seen in different stages of differentiation. In a few of the largest cells, the parasite may have subdivided into many minute rounded spores. These spores are released into the soil when the root decays, and remain infective to attack other susceptible roots.

3.2.2 Water-moulds and related fungi

Most of the lower Phycomycetes are saprophytic water-moulds, but an exception is *Saprolegnia parasitica* which parasitizes fish in fresh water. It is often possible to find in streams dying sticklebacks and minnows covered with a weft of aseptate mycelium of this fungus. The only terrestrial representatives of these fungi are several species which parasitize plant roots, such as *Synchytrium endobioticum* which causes a wart disease of potato tubers and *Aphanomyces* spp. on several crop plants. These fungi are released from decaying plant roots to survive in soil as dormant spores.

3.2.3   Peronosporales

Also among the Phycomycetes, the order Peronosporales contains three families of important plant parasites, Peronosporaceae, Albuginaceae and Pythiaceae. These families are distinguished mycologically by the morphology of their asexual spore-bearing structures, but they are also distinct by the symptoms of the diseases which they cause.

The Peronosporaceae cause the downy mildew diseases, which appear as soft white threads on the surfaces of infected stems and leaves. The threads are sporangiophores which grow out through the stomata from infected tissues. Sporangiophores should be removed gently to a drop of water on a glass slide, and covered carefully with a cover slip before microscopic examination. The structure of the sporangia and the way in which they are borne on the sporangiophores should be seen without difficulty. These features permit identification of the genus of the parasite as indicated in Fig. 3–1. Genera and species of these obligate parasites are confined in their ability to attack certain species of host plant. *Bremia lactucae* parasitizes lettuce, *Peronospora destructor* onion, *Plasmopara viticola* vine and *Pseudoperonospora humuli* hops.

Because these are obligate parasites, the only method of culture is to spray a batch of plants of an appropriate susceptible species with a suspension of freshly harvested sporangia in water, and then to incubate the plants at about 20°C for several days beneath a polythene hood over a tray of water, to maintain high humidity. A new crop of sporangiophores should develop on the surfaces of the plants.

The members of the Albuginaceae are obligate parasites, and cause white blister rusts on a number of plants. Well-known and easily found in summer months is *Albugo candida* (*Cystopus candidus*), which causes white blisters on members of the Cruciferae. It can always be found on the weed, shepherd's purse, *Capsella bursa-pastoris*. Specialized physiologic races of the parasite cause considerable damage to crops, such as cabbage or radish. Blisters are areas of raised host epidermis over fungal pustules in which sporangia are cut off in chains from tips of club-shaped sporangiophores, as can easily be seen in sections viewed under a microscope. The parasite is spread as sporangia in wind or water, but overwinters either as dormant oospores, which have thick, warty walls, or in perennial plants as dormant mycelium.

The Pythiaceae are distinguished from the other two families by the fact that they produce sporangia on unspecialized hyphae. The family contains two important genera, *Pythium* and *Phytophthora*, which are mostly soil-inhabitants. These fungi can be selectively isolated from soil with the aid of unripe apples, which permit internal growth of these fungi but not of others. One side of the apple should be punctured with a needle, and then pressed into a soil sample. After a few days in a moist environment, areas of brown rot may be found in the fruit. The fungi may be grown out from pieces of rotted fruit, handled aseptically, on to culture media. Species of *Pythium*, which cause damping-off diseases of seedlings,

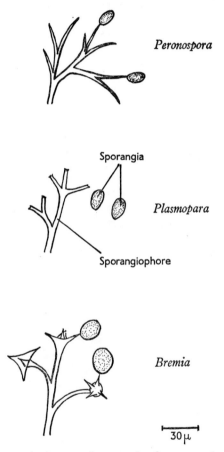

Fig. 3-1   Sporangiophores and sporangia of common downy mildew fungi.

can be obtained by germinating cress in soil samples in polythene bags. *Pythium* grows very rapidly on media providing one-tenth the nutrient concentrations given in Chapter 2. Some of these readily grown parasitic species of *Pythium* have special biochemical needs for spore production. These needs may be met by natural substrates such as oat meal or sterilized grass leaves, or by supplying certain sterols.

The genus *Phytophthora* contains important root-infecting parasites, which are particularly troublesome in places where intensive agriculture is aided by canal irrigation as in California. Motile flagellate zoospores spread the parasites in water films on soil particles. The leaf parasite, *Phytophthora infestans*, greatly influenced the history of the world through the late-blight disease of potatoes. The disease became serious throughout Europe between 1840 and 1850, and caused the Irish potato famine in 1845. Much of the Irish population died, or emigrated to the U.S.A. in particular. *P. infestans* can overwinter as mycelium in potato tubers, and grow into young shoots in spring. Sporangia are rapidly produced on aerial parts in warm wet weather. Sporangia are spread by wind and rain-splash, but die in dry conditions or over 20°C. Under the common conditions of British and Irish summers, potato plants are readily infected. Very rapid epidemics of late-blight develop in 2 day periods of 90–100% relative humidity and temperatures of 15–20°C. At present, farmers are warned when the blight fungus appears and when the weather is likely to favour the disease. Fungicidal sprays can then partially protect the foliage. Serious epidemics still occur, however. The extreme step is sometimes necessary of killing both foliage and fungus by spraying with sulphuric acid several days before harvesting tubers. Otherwise tubers may be infected through small wounds during harvest, and infected tubers rot in store.

3.3 Ascomycetes

3.3.1   Taphrinales

All members of this order are parasitic, and cause growth deformations in leaves and 'witches' brooms' in branches of trees. Particularly common are the species of *Taphrina* which cause leaf curl in peaches and cherries. Strongly puckered and reddened diseased leaves are easily found from June onwards as in Plate 11. Thin sections through these leaves should reveal single layers of sac-like asci under ruptured leaf cuticles. Although the fungus is mycelial in plants, it grows by budding like yeasts in culture.

3.3.2   Erysiphales

These very important obligate parasites cause the powdery mildew diseases named after the dry and mealy appearance of the mycelium and conidia on plant surfaces. The mycelium grows outside the cuticle but is anchored to the plant by haustoria pushed into epidermal cells. Oval, single-celled conidia are produced from simple conidiophores. Sexually-produced fruiting bodies (cleistothecia) can be found in late summer and autumn on most powdery mildews. Powdery mildews overwinter as cleistothecia or resting mycelium. Mature cleistothecia are black, spherical bodies visible to the naked eye, and can be lifted from the mycelial mat and mounted on microscope slides. Hyphal appendages should be seen

clearly under low-power magnification. Asci can be forced out of cleistothecia, while they are being viewed, by gentle pressure on the cover slip
with a dissecting needle. The structure of the appendages and number of
asci are used to distinguish powdery mildews, as indicated in Fig. 3–2.

Some powdery mildews have wide host ranges, such as *Erysiphe polygoni*

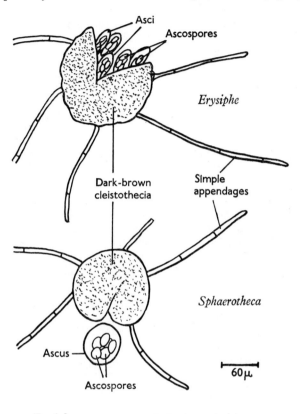

Erysiphe   – many asci, simple appendages
Sphaerotheca – one ascus, simple appendages
Podosphaera – one ascus, appendages with
                        forked tips

Fig. 3–2 Burst cleistothecia of powdery mildew fungi.

on many plants including pea and species of *Brassica*, and *E. graminis* on
grasses. Many physiologic forms of *E. graminis* can be recognized by
testing abilities of isolates to parasitize different species and varieties of
grasses. Other powdery mildews have confined host ranges, for example
Podosphaera leucotricha on apple and *Sphaerotheca mors-uvae* on goose-

berry and blackcurrant. All of these are easily found in the summer, but perhaps the most common is *Sphaerotheca pannosa* on roses.

It is interesting to note the weather conditions and positions in gardens where powdery mildews are prominent. Rose mildew will occur on some varieties only, and particularly on roses against houses where the bushes are sheltered from rainfall, dew formation and air circulation. Many experiments have been done to attempt to understand the distribution of powdery mildews. Unlike most other conidia, those of powdery mildews have high water contents. Although germination is best in humid atmospheres, liquid water can be inhibitory and sometimes lethal to the conidia.

Some powdery mildews do little more than disfigure their host plants, but others such as *Uncinula necator* on grape vines can destroy whole crops. In general, infections can be minimized by encouraging air circulation through susceptible plants by, for example, thinning the centres of rose bushes. Disease can be controlled by sprays with lime sulphur (aqueous calcium polysulphides) or some organic fungicides sold under brand names.

3.3.3 Other Pyrenomycetes and Discomycetes

The majority of these fungi are saprophytic, but most groups have a few parasitic representatives. *Endothia parasitica* is one of the most destructive; it was introduced into North America in the early twentieth century and has now eliminated the American chestnut from that continent.

Claviceps, the ergot fungus, is an interesting parasitic Pyrenomycete, and has several important implications for man. The form of the parasite which is most easily found is the sclerotium or ergot. The hard black sclerotia are formed in place of grains in infected grass spikelets (see Plate 4). At harvest, sclerotia are knocked to the ground, where they overwinter and germinate in the following spring. Small mushroom-like growths emerge from the sclerotia and bear embedded perithecia containing asci. Mature ascospores are forced out into air currents at the time when grasses are flowering. On the flowers of susceptible grasses, ascospores germinate and infect ovaries. Masses of mycelium produce many small conidia in a sugary secretion, which attracts insects. Insects transmit conidia to other flowers. The mycelium which replaces ovaries hardens and darkens to form sclerotia, which are easily seen because they protrude from the grass spikelets. The main economic effect of ergot is not that grain yields are seriously depleted but that the sclerotia contain poisonous alkaloids. These alkaloids when processed in flour and bread cause ergotism in man. Cattle which graze infected grass can become ill and may die. The ergot fungus is used industrially to produce some alkaloids of medicinal value. Many species of *Claviceps* occur on different grass hosts, and the related genus *Cordyceps* is parasitic on insects.

A common parasitic Discomycete is *Rhytisma acerinum*, the cause of the so-called tar spots on sycamore leaves. The tar spots are masses of black hardened hyphae, and overwinter in leaf debris. They split in spring to expose apothecia which eject clouds of ascospores into air currents. Ascospores produce germ-tubes and infect young *Acer* leaves via stomata, eventually forming new tar spots. The parasite *Sclerotinia fructigena* referred to in Chapter 2 as the cause of brown rot of apple fruit is a Discomycete. This and related species cause very important diseases of fruit in many parts of the world.

3.3.4 Ascostromatic fungi

These fungi bear their asci in cavities in the stroma of their fruiting bodies, and not in true perithecia. Two genera are particularly important plant parasites, namely *Mycosphaerella* and *Venturia*.

Mycosphaerella musicola causes a destructive leaf-spot of bananas, the Sigatoka disease, and forces the installation of costly pumping and spraying machinery in banana plantations for its control. Other species of *Mycosphaerella* cause important leaf-spot diseases, in pea and strawberry for example.

Species of *Venturia* cause serious diseases of leaves and fruit of apple and pear (see Plate 5), and the most common is *V. inaequalis*, which induces the formation of brown patches in apple leaves, and of scabs in fruit. This parasite overwinters in leaf debris. Under the influence of spring rains and dews, ascospores are ejected. After dispersal in the wind, ascospores germinate in water droplets and infect young susceptible leaves. In wet weather the fungus grows rapidly beneath the cuticle and eventually produces conidia through burst cuticles. Conidia are dispersed to other leaves so that in favourable wet conditions apple scab epidemics build up. Premature leaf fall can follow heavy infection. Scabs form on young fruit and greatly reduce their market value. In autumn, mycelium grows deeply into leaves and forms the overwintering stromatic tissue. Pairing between nuclei from opposite mating strains of the parasite precedes ascus formation in the stroma. The slow maturation of the asci continues into the spring, and is influenced by winter temperatures and rainfall. Because leaf litter is the source of *Venturia* in the spring, cleanliness of the orchard floor is a major factor in disease control. Eradicant chemicals can be applied to the leaf litter in autumn and winter. Natural control occurs by removal of fallen leaves, by microbial action and by earthworms. During spring and summer, trees are sprayed with fungicides at times recommended by plant pathologists working with meteorologists. It is desirable to collate their advice to minimize fungal proliferation and infection in summer and to eradicate resting mycelium in winter.

3.4 Deuteromycetes

Many fungi which are usually found as mycelia or asexual reproductive spores are grouped for convenience of identification in the Deuteromycetes.

Some of these fungi are known Ascomycetes, others are very similar to Ascomycetes in their asexual structures and some may be Basidiomycetes. A few of the common and important parasites from the many thousands of Deuteromycetes are described below.

3.4.1   Sphaeropsidales

The characteristic of these fungi is the production of dark, flask-shaped bodies (pycnidia) containing masses of conidia. Pycnidia can be teased from leaves and stems of host plants, and examined microscopically. Pycnidia can be encouraged to exude their conidia if the plant bearing them is incubated on moist filter paper in a plastic box for a few days. The shape and size of the conidia permit identification of the fungus.

Long, slender, septate conidia are produced by the fungi comprising the form-genus *Septoria*. Some species of *Septoria* produce sexual stages at certain times of the year, and can be classified as Ascomycetes. *Septoria rubi*, the cause of raspberry leaf spot, and *S. avenae*, the cause of oat leaf spot, are then found to belong to two widely separated genera of Ascomycetes. The convenient grouping of fungi by their asexual structures often causes similar mistakes about their affinity.

Another recurrent problem in the classification of parasitic fungi is exemplified by the description of over a thousand species of *Septoria*. These have been described, not usually because of morphological differences, but because a new species has been claimed each time *Septoria* has been found on a new host. Some of these species have been re-examined by isolating them from their hosts, comparing their morphology on standard culture media, inoculating them on to a range of test plants and checking their morphology on their new hosts. These laborious procedures have enabled regrouping into fewer distinct species.

3.4.2   Moniliales

In these fungi, the conidia are borne freely on the mycelium and not in specialized organs. Among the Moniliales are important human parasites, plant parasites and many saprophytes. Some of the human parasites affect the lungs causing diseases such as aspergillosis. Others affect the skin causing diseases known as dermatomycoses, which include athletes' foot.

The colour and structure of the conidia and conidiophores are the features used to identify these fungi. Intact conidiophores are not easily transferred to microscope slides, so a slide culture technique can be used to produce conidiophores for direct observation. A 1 cm^2 block of agar is cut from medium in a petri dish and placed on a sterile microscope slide. A small amount of inoculum is placed on the block and covered with a sterile cover slip. After incubation of the slide for a few days over moist filter paper in a petri dish, direct observation can be made of the ways in which conidia are produced and carried on conidiophores. If necessary the cover slip plus attached conidiophores can be transferred to a drop of cotton blue in lactophenol on a new slide, before examination (see Fig. 3–3 and 3–4).

Most easily studied parasitic Moniliales are *Penicillium* and *Botrytis* species as discussed in Chapter 2. Common parasitic *Penicillium* species are *P. digitatum* and *P. italicum,* the causes of green and blue mould of citrus fruits respectively, and *P. expansum,* which rots apples. The form-genus *Botrytis* contains many species some of which may be amenable to regrouping as *B. cinerea* using the cross-inoculation tests described for *Septoria.* Conidia of *Botrytis* can be found in most gardens in wet weather on many flowerheads, bean pods, onions, lettuces and on soft fruit, such as strawberry.

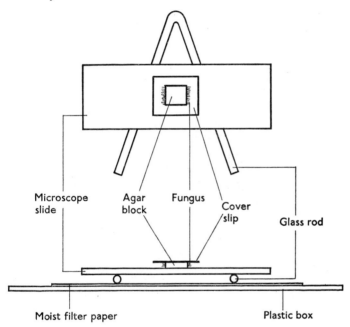

Fig. 3–3   Slide culture technique.

The most important group of plant parasites in the Moniliales is the form-genus *Fusarium.* Most parasitic species of *Fusarium* occur in soil, often as resting chlamydospores, and they attack the roots of susceptible plants. Commonly, most fungal growth is made in the xylem vessels of the vascular system in the root and the base of the stem. By mechanisms which are not fully understood, the fungi cause wilt of the leaves and often death of the plant.

These wilt diseases are common and serious, largely because infected soil is rendered useless for certain crops for many years. The most serious is

Panama wilt of bananas caused by *Fusarium oxysporum* f. *cubense* in Central America. Pea and tomato wilts are important in Britain, W. Europe and N. America. Because of their economic importance, the correct identification of isolates of *Fusarium* is desirable but difficult, even for specialists. The variability between isolates is great, and is much affected by culture media and environment.

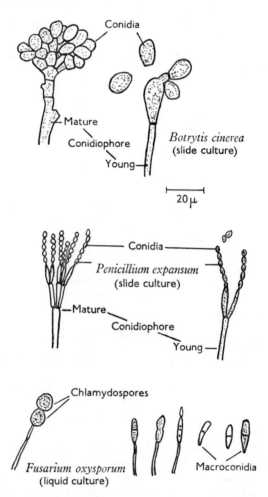

Fig. 3–4   Conidia of some common fungi in the Moniliales.

3·5   Basidiomycetes

Three groups of parasitic fungi must be considered under this heading, rusts, smuts and some of the conspicuous types, which everybody recognizes as fungi when they find fructifications on trees and fallen timber in woods and forests.

3.5.1 Uredinales—the rusts

The rusts are important and well-known because they have conspicuously affected man's crops throughout recorded history; they are obligate parasites with a succession of spore types, often on two alternate species of host plant.

The rusts are regarded as highly specialized parasites for three reasons. First it has proved impossible to culture these fungi on anything but their normal host plants. Secondly, many rusts are only able to parasitize one species of host. Thirdly, some rusts exist as many physiologic races, each of which can only parasitize a single host variety. In fact success or failure of parasitism on a particular variety may depend upon single genes in parasite and host.

The maximum number of different spore types in any particular rust is five, as possessed by the parasitic fungus most often described in text books, *Puccinia graminis tritici*, black stem rust of wheat. Uninucleate spermatia and binucleate aecidiospores are formed on barberry and binucleate uredospores and teleutospores on wheat. The teleutospores germinate and give rise to uninucleate basidiospores. Nuclear transfer between mating strains of rust occurs during the spermatial stage in the barberry leaf. This rust is very important economically, and limits wheat production in the world's greatest wheat growing area, the central regions of the U.S.A. and Canada. The alternate host, barberry, was eradicated by plant pathologists in this region, but the disease on wheat continues without the spore types on barberry. Uredospores are carried in northward moving air masses each spring from wheat overwintering in the south and Mexico. As polar air masses encroach southwards from Canada in late summer, uredospores can be carried back from the prairie states. The main control measure at present is the breeding and use of varieties of wheat resistant to the predominant physiologic races of the rust. There is a constant threat, occasionally realized, of a new race of rust evolving, by genetic recombination or mutation, to parasitize current wheat varieties.

Some common spermatial and aecidial stages of rusts are shown in Fig. 3–5 and 3–6 and Plate 6. Even more easily found in summer months are the uredial and telial stages of rusts as illustrated in Fig. 3–7 and Plate 12. The main feature of the genus *Puccinia* is that the teleutospores are two-celled. Common species in gardens are *Puccinia malvacearum*, which bears teleutospores on hollyhock, and *P. antirrhini*, which produces uredospores and teleutospores on *Antirrhinum*. The other large genus of rusts is *Uromyces*, which has single-celled teleutospores. A common species, *Uromyces viciae-fabae*, produces all five spore types on broad bean and wild species of *Vicia* and *Lathyrus*.

22

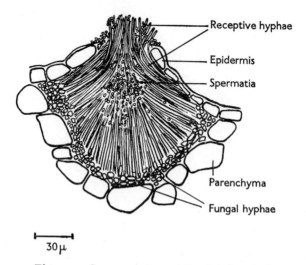

Receptive hyphae

Epidermis

Spermatia

Parenchyma

Fungal hyphae

30μ

Fig. 3-5 Spermogonium of *Puccinia* in a leaf.

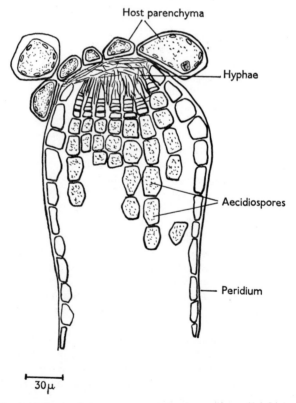

Host parenchyma

Hyphae

Aecidiospores

Peridium

30μ

Fig. 3-6 Aecidium of *Uromyces* projecting through lower epidermis of leaf.

Other economically important rusts are *Hemileia vastatrix*, the ruin of the coffee exporting business in Ceylon in the nineteenth century, and *Cronartium ribicola*, the white pine blister rust. Damage to white pine in N. America is being minimized now at the expense of currants and gooseberries, which, as the alternate hosts, are being eradicated by plant pathologists.

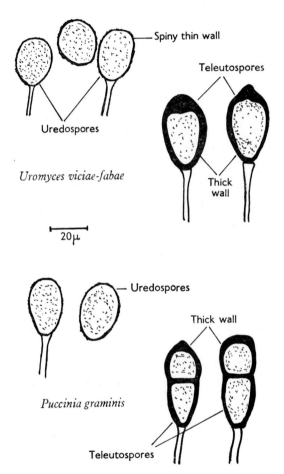

Fig. 3-7 Uredospores and teleutospores of common species of *Uromyces* and *Puccinia*.

3.5.2   *Ustilaginales—the smuts*

Smuts are highly specialized parasites, and typically transform specific parts of infected plants into dark brown masses of dusty spores. Smuts are not obligate parasites in a complete sense, because they can be grown on culture media although in an atypical yeast-like manner.

Many smuts parasitize grasses, and transform the grains into masses of spores. Grains of the cereal grasses can be found in this state at harvest time. A common example is loose smut of wheat, *Ustilago tritici*. The smut spores are carried in the wind, and those landing in suitably moist places germinate by small hyphae. Further development is dependent upon two chance events. First, the germination hyphae must be so close to the hyphae from another mating strain of smut that nuclear transfer can occur between compatible strains which fuse. Secondly, the resulting mycelium now with two nuclei per cell must be on the flowers of a susceptible host. If so, the mycelium grows into the ovary and lies dormant in the developing seed. If the mycelium has reached the growing point of the embryo in the germinating seed, it will be carried up the stem passively as the seedling elongates. The mycelium only starts to grow rapidly in the flower initials. This is how the grains are replaced by a mycelial mass, which divides to form many smut spores. These spores are equivalent to rust teleutospores. Some typical spores are shown in Fig. 3–8.

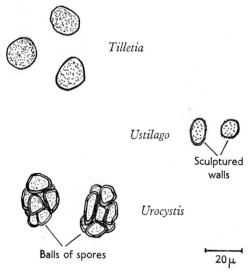

Tilletia

Ustilago

Sculptured walls

Urocystis

Balls of spores

20 μ

Fig. 3–8   Some smut spores and spore balls.

Each of the many different smuts occurs on a narrow range of hosts, and affects a particular plant part (see Plate 7). *Urocystis cepulae* erupts as spore masses through infected onion leaves. Spores of *Ustilago violacea* replace the pollen from anthers of red campion and related plants. The most important smuts economically are those on the cereals, where avoidance by using healthy seed is the best control measure.

3.5.3   The large Basidiomycetes

The familiar encrustations, brackets, toadstools and similar fructifications found on decaying timber are fruiting bodies which bear basidiospores. Hidden from view is the colonizing and feeding mycelium, which
permeates wood and other organic matter. Many fungi colonize fallen
timber as saprophytes. Others colonize living trees, but not parasitically,
because they enter through wounds and grow in the central dead heartwood.
Relatively few of the higher Basidiomycetes can be termed parasites with
confidence.

Stereum purpureum is an undoubted tree parasite which causes silver-
leaf diseases in plum trees as well as invading freshly fallen timber where
the parenchyma is still alive. The numerous small crusty fructifications
are grey and hairy above, and have purple sporing surfaces. The fungus
enters branches of fruit trees through pruning wounds, and grows in
living tissues. The silver-leaf effect is caused by the separation of tissues
and the creation of air spaces in the leaves. The fungus produces substances which are translocated to the leaves where they cause cells to
separate.

More questionable parasites are the larger bracket fungi on standing
trees, such as *Polyporus betulinus* on silver birch, *P. squamosus* on elm in
particular and *P. sulphureus* on oak. These fungi enter through wounds,
and cause heart rots by growing through and degrading the heartwood.
The sudden fall of apparently healthy branches of elm is a result of heart
rot.

Similar questions of extent of parasitism are posed by *Ganoderma
applanatum*, the common cause of heart rot in beech. This fungus enters
via wounds in large trees after wind or lightning damage. The perennial
fructifications, producing new sporing surfaces each year, are shelf-like and
dusted with rust-colored spores.

An undoubted parasite among the toadstool type of fungi is *Armillaria
mellea*, the honey fungus. This parasitizes many trees and some shrubs
in temperate and tropical regions of the world. Characteristic structures
are black rhizomorphs found beneath the bark of infected trees. Rhizomorphs are strands of hyphae, hardened and darkened. The yellow-brown
toadstools have brown scales on their upper surfaces and white gills under-
neath, bearing colourless spores. Sometimes the fungus is found growing
saprophytically on decaying matter and not affecting neighbouring trees.
It is likely that specialized parasitic and saprophytic strains of the fungus
exist.

Some of the large Basidiomycetes live on roots of forest trees, but they
also confer some benefits to the roots so that they are said to have symbiotic
and not parasitic relationships with the trees. Particularly important sym-
bionts are species of *Boletus*, known as ceps or bun-fungi, because of their
soft bun-like stalked fructifications with porous spore-bearing surfaces

beneath. Each species of *Boletus* is associated with a particular tree. The symbiotic association of mycelia and rootlets form mycorrhizal roots, which can be found beneath leaf litter in forest soils, as distinctly white or yellow strands.

Spread of Parasitic Fungi 4

Parasitic fungi have specialized ways of getting from one host to the next.

4.1 In soil

Fungal spores liberated into soil may move slightly in water films on soil particles. Most root-parasitic fungi spread slowly. They rely on the active growth of roots into their vicinity to supply new host tissues. For example, *Plasmodiophora brassicae* does not spread unless man moves it in infected plants or seed. It remains viable as resting spores in soil for many years. Sufficient spores remain ungerminated each season to await the arrival of susceptible roots of cruciferous plants when these are grown again after a fallow period or crop rotation. Mobile zoospores of the fungus spread over the short distances separating resting spores from the roots which stimulated their germination.

Exceptional among parasites of roots for their rapid spread are some species of *Phytophthora* in parts of the world where canals are used to irrigate crops. Citrus plantations in southern California are badly affected in this way by *P. citrophthora*. The existence of the parasite in the irrigation canals was shown by trapping the fungus in lemons held in perforated plastic bags suspended in the water. Zoospores infected the fruit, caused brown rot and were isolated in the laboratory. Irrigation canals become contaminated by water draining from infected citrus groves.

The spread of many root parasites can be checked by quarantine regulations forbidding import of infected seed, plants and soil. The parasites can then sometimes be eradicated by applying fungicidal treatments to the soil and by not planting susceptible roots for many years.

4.2 From aerial parts of plants

Wind is the major factor in the spread of parasites of leaves and stems, locally through crops and distantly over quarantine barriers to new areas. Aerial spread depends upon some properties of the atmosphere and special features of the parasites.

4.2.1 Properties of the atmosphere

Several different layers in the atmosphere can be defined between the surfaces of solid objects and the stratosphere where the air is beyond the influence of the earth. The layers shown in Fig. 4–1 vary in thickness and precise limitation with the conditions. Over all surfaces is a layer of still air. This very thin layer is covered by a layer of streamlined air which

varies in thickness with the wind speed and with surface roughness.
Minor projections such as hairs on a leaf cause turbulence in flowing air.
At very low wind speeds, small projections may be covered with still air
so that they do not cause turbulence. Major projections such as trees and
buildings cause major turbulence, particularly in strong winds. In turbulent
air particles can be carried rapidly upwards away from surfaces. Major
upward currents of air rise above land warmed by the sun. The most
obvious consequence is the formation of cumulo-nimbus cloud when cool
air flows over moist warm air. This moist warm air rises rapidly and forms
the thick towering clouds typical of thunderstorms. Particles can be carried
to altitudes of many thousands of feet in these thermal upcurrents.

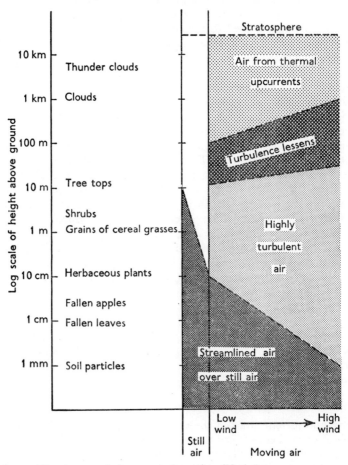

Fig. 4-1 The layers of the atmosphere (modified from GREGORY, P. H.,
1961; see Further Reading).

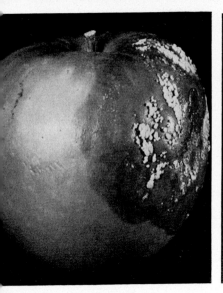

Plate 1 Brown rot of apple. Several days after wound inoculation, *Sclerotinia fructigena* is producing spores on the white pustules in the brown rotted tissue (from Dr. B. E. J. WHEELER, Imperial College).

Plate 2 Green mould of orange. Several days after wound inoculation, *Penicillium digitatum* is producing spores (dark patches) on the white mycelium covering the rotted tissue.

Plate 3 Chocolate spot development on detached leaves of broad bean in plastic sandwich box. On the left half of each leaf, *Botrytis fabae* is causing black spreading lesions. On the right half of each leaf, *Botrytis cinerea* has caused brown spots which remain limited.

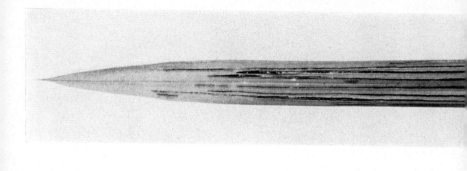

Plate 7 Striped smut on leaf of *Glyceria maxima*.

Plate 6 Aecidia of rust, *Uromyces dactylidis*, on leaf of *Ranunculus*.

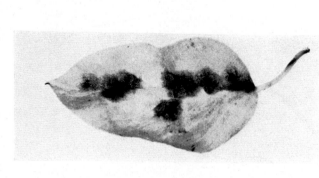

Plate 5 Scab lesions caused by *Venturia pyrina*, in young pear leaf.

Plate 4 Ergots of *Claviceps purpurea* in spikelet of rye (from Dr.

Air masses have characteristic movements around the world. Heating by the sun causes air to rise over the Equator, and to different extents over oceans and continents. Air descends over the two Poles. Friction between the rotating Earth and the air causes lateral flow of air masses. Therefore complex patterns of air circulation occur around the world as shown in maps of trade winds. Polar and tropical, and marine and continental air masses can be recognized.

4.2.2   Take-off into the wind

To be dispersed in moving air, fungal spores must escape from surface layers of still air. Some fungal spores are carried into the air, others are forcibly ejected. Moving air can dislodge conidia from aerial conidiophores, and carry the spores away. Droplets of water as mist or rain-splash can carry spores into the air and liberate the spores as they evaporate. Most of the active methods used by fungi to eject spores into the air are based on rapid changes in water content in the hyphae. Some conidiophores twist violently and throw off their conidia. Mature asci of *Venturia* and apothecia of *Rhytisma* puff out ascospores into the air. Basidiospores are shot off from the specialized short hyphae which bear them.

4.2.3   Dispersal in the wind

Success in escaping from surfaces enables spores to be swept away in moving air. The distances over which spores are dispersed depend upon several factors. Observe smoke being dispersed from chimneys in different weather conditions. In still weather the smoke either rises vertically into upper cold air, or falls back over the ground if air temperatures are higher than those on the ground. In windy weather the plume of smoke widens and becomes less dense as it blows away. Masses of spores are carried away in the same way.

All particles in the air tend to fall back to earth under the influence of gravity. Their speed of fall in still air depends upon their diameter and density. Densities of fungal spores are approximately similar to that of water. Relative rates of fall of different spores depend mostly on their sizes. Rates of fall can be found by measuring the time required for a dispersion of spores to reach the bottom of a settling tower. A simple tower might be a measuring cylinder. Complicating air movements in the tower can be caused by convection currents from local sources of heat. Such problems might be minimized by using a wooden tower in a room away from beams of sunlight, radiators and draughts. The tower should provide a means of adding spores at the top and of inserting an open petri dish at the bottom. Dishes should be carefully exchanged at intervals without unduly disturbing the air in the tower. Microscopic examination of the dishes should reveal the presence of spores. Suitable dispersions of spores are those of *Lycopodium*, club-moss (about 30 μ diameter), obtainable from British Drug Houses Ltd. for example, rust uredospores (about 20 μ diameter) and basidiospores of bracket fungi (3–10 μ diameter) collected from fields and woods. It will be found that clumps of spores fall more quickly than single spores. Typical rates of fall for single spores are 2 cm/sec for *Lycopodium*, 1 cm/sec for *Puccinia* and about 0·1 cm/sec for the smaller basidiospores.

The fate of spores freed into the wind depends very much on wind-speed and presence of thermal upcurrents in air. As air movement slackens, the dominant force on spores becomes that of gravity. Obviously small spores have greater chances of distant dispersal than large spores.

4.2.4   Spores in the air

The dispersion of spores in the air makes up the air spora. The spora can be sampled by catching spores on sticky surfaces. A simple method is to expose surfaces of agar in petri dishes horizontally or vertically in the air. The exposed surfaces can be examined microscopically to reveal pollens, spores and other particles. Some of the spores trapped on culture media will produce fungal colonies if the dishes are incubated at 25°C for several days. Markedly different collections of particles will be obtained at different times in different environments such as city centres, parks and near farms.

Catching spores on exposed sticky surfaces is much less efficient than may be imagined. A major trouble is that, as air flows towards the surface, turbulence is created and many particles are carried around the trap.

Table 1   Densities of spores trapped in the same place on slides in a volumetric trap and exposed (modified from HIRST, J. M., 1959, in *Plant Pathology Problems and Progress, 1908–1958*; see Further Reading).

Fungal spores	Average numbers of spores/cm²	
	Hirst spore trap	Horizontal slide
Smuts	621	3
Alternaria	156	3
Cladosporium	8930	59
Erysiphe	100	2

Large particles will be caught more readily than small particles which will be deflected more easily in the turbulent air. For similar reasons efficiency of trapping will vary with wind-speed.

Many problems of efficiency of trapping have been reduced by use of mechanized traps such as the Hirst volumetric spore trap. This trap is directed to face the wind by a wind vane. Air is sucked through the trap at a known speed by a vacuum pump. Air enters through a slit and passes over a microscope slide coated with adhesive where spores are impacted.

The slide is drawn slowly past the slit by a motor so that a continuous record of trapped particles is obtained over a 24 h period. This trap is independent of wind speed and direction, and it captures most particles in a known volume of air. Its great catching power compared with that of an exposed horizontal slide is shown by Table 1.

Volumetric spore traps are used by plant pathologists and by medical teams working on the relationship between spores, pollen and allergies such as asthma. Effects of weather, season and time of day on air spora can be found with great accuracy.

Spore trapping from aircraft has shown that the major air masses moving around the world carry characteristic air spora. Spores are found high over the Atlantic and Arctic. The composition of the Arctic air spora shows that it can be derived from temperate regions far to the south. The aerial movement of uredospores of cereal rusts northwards over the U.S.A. and Western Europe has been deduced not only from outbreaks of disease but also from samples of the sky above.

4.2.5   Return of spores to the ground

In still air, spores fall to surfaces on earth under the influence of gravity. Air is rarely still, and various specialized ways for the return of spores to surfaces exist. One important method is impaction of spores on to projecting objects in moving air, as revealed by experiments using wind tunnels. Particles in motion in air tend to move in straight lines unless they are deflected by air moving in other directions. Larger particles maintain their straight courses more readily than smaller ones. These straight courses may be towards aerial parts of plants, where the spores impact. Efficiency of impaction increases with wind speed, and as targets become smaller causing less deflection of air. In effect this means that the very small spores of soil-inhabiting fungi such as *Penicillium* would rarely impact on any object. Spores of the smut, *Ustilago*, are about 8 μ diameter and only impact efficiently on small projections such as parts of the flowers of grasses. Larger spores such as rust uredospores and powdery mildew conidia impact efficiently on narrow stems and leaves. Very large spores, 90 μ diameter, of *Botrytis polyblastis* would impact very efficiently on thick stems, such as those of its host, daffodil.

Turbulence created by wind flowing around a leaf may cause spores to impact on the under-surface of the leaf. Some leaf rusts are more commonly found on lower sides of leaves, and this may be caused in part by the deposition of spores.

The other major way in which spores return to plant surfaces is by rain-washing of the air spora. After heavy rain, very few spores may be found in the air for a period. The efficiency of rain-washing depends upon sizes of raindrops and of spores. Large spores are collected easily by all sizes of rain droplet, but spores less than 8 μ diameter are brought down only by the larger raindrops.

4.2.6   The effects of wind dispersal

The potential effects of wind dispersal of parasitic fungi are most simply shown by the spread of parasites into parts of the world where they did not previously exist. It is almost impossible to discover the extent of dispersal within already widely distributed parasitic species.

Peronospora tabacina causes the destructive blue-mould disease of tobacco, and has been known in eastern Australia since 1890. Man introduced it accidentally into England in 1958. Its further dramatic spread into Europe, Africa and Asia was caused by its natural method of wind dispersal. In 1959 it appeared on tobacco in Holland and Germany and in 1960 it occurred throughout Europe destroying about 65% of the crop.

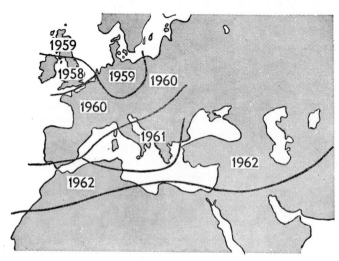

Fig. 4–2   Spread of *Peronospora tabacina* on tobacco crops in Europe (modified from GREGORY, P. H., 1963, *Advancement of Science*, March, 481–488).

In 1961 it was in the Mediterranean, and in 1962 in N. Africa and the Middle East, as shown in Fig. 4–2. Its spread is only likely to be limited by vast areas lacking tobacco, such as deserts and oceans.

Dispersal of other small-spored parasites of leaves and stems is likely to be equally extensive and rapid. The consequences are particularly important for man where new parasites are involved, such as new physiologic races of widespread parasites.

4.3   Rain-splash dispersal

Apart from accidental carriage of fungal spores by animals, man and his machines, the other major dispersal method is by splash droplets during rain. This results in local spread of parasitic fungi. When raindrops fall on to surfaces coated with masses of sporophores many spores are picked up in the droplets which splash away in all directions. Large splash droplets pick up more spores than the small ones which may become airborne. The large droplets follow trajectory courses, and deposit their contained spores on surfaces which they strike. A combination of wind and rain-splash dispersal causes most of the spread of parasitic fungi in stands of vegetation. Late blight spreads through a potato crop in this way.

Three phases can be distinguished in the successful colonization of a plant by a parasitic fungus. The first occurs when plant and fungus become sufficiently close for the plant to stimulate activity in the fungus. Then the fungus enters the plant by a specialized route. Finally the fungus grows between and into the plant cells, often causing their death through the action of enzymes and toxins which it secretes.

5.1 Pre-penetration events

As a consequence of their mode of dispersal, many spores land on aerial parts of plants and others lie as resting stages in the soil. Spores of some parasitic fungi germinate in water alone, but many spores, particularly resting spores, must be stimulated to germinate by physical and chemical factors. The most important influences of plants on fungal spores are as sources of chemical stimulants and inhibitors.

5.1.1   *In soil*

Plant-parasitic fungi rarely have free existence in soil, but occur as resting spores or sclerotia. They germinate when conditions become suitable for their parasitic activities. Not only must temperature and soil moisture contents be favourable for germination but the resting structures must be stimulated by plant roots. The dormant state of spores in soil is known as fungistasis. Fungistasis is probably caused by antibiotic factors produced by other soil inhabitants, particularly saprophytic fungi and bacteria. Fungistasis is overcome by low concentrations of substances exuded from living roots.

It is relatively simple to show that sugars and amino acids are exuded from germinating seeds and roots. Pea or bean seeds are germinated on moist filter paper. At different stages during the growth of seedlings, sample papers are air dried and dipped into location reagents. A freshly mixed solution of $1\cdot3$ ml aniline in 50 ml acetone and $0\cdot6$ ml H_3PO_4 in 20 ml acetic acid plus 30 ml acetone stains sugars, like glucose, yellow-brown after evaporation of acetone and heating at $100°C$ for 2–3 min. A solution of $0\cdot1$ g ninhydrin in 50 ml acetone stains amino acids blue-purple after drying and heating at $105°C$ for 5 min. Particularly intense colouration will appear on those parts of the paper corresponding with the micropyle of the seed and with rapidly growing areas behind root tips. See the diagram in Fig. 5–1.

The stimulatory effects of solutions of sugars and amino acids (e.g. 1 mg/ml) can be demonstrated using the spore germination test described earlier. Water which has bathed germinating seeds can be shown to stimulate spore germination and germ-tube growth in the same way. A necessary precaution is to avoid using seeds which have been treated with fungicides.

Such exudates stimulate microbiological activity around the roots in what is known as the rhizosphere. Resting spores of Phycomycetes release zoospores which swim along concentration gradients of exudates to the root surface, where they encyst and produce penetration hyphae. Resting

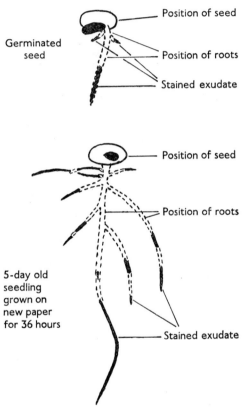

Fig. 5–1 Pattern of exudation of sugars and amino acids from a young seedling (redrawn from SCHROTH, M. and SNYDER, W. C., 1961, *Phytopathology* **51**, 389–393).

spores of non-motile fungi produce germ-tubes which grow towards the sources of nutrients and on to the root surface.

Most roots stimulate the growth of most fungi, both saprophytes and parasites. There is little evidence that different roots selectively cause the germination of their specialized parasites, but two interesting observations can be cited. Motile zoospores of *Phytophthora cinnamomi* are attracted to their host roots, avocado, but not to those of citrus. Conversely, zoospores of *P. citrophthora* are attracted to their host roots, citrus, but not to those of avocado. There is as yet no explanation of this specific response, and most attempts to find similar phenomena with other root parasites have given negative results.

5.1.2   On aerial parts of plants

Most spores which fall on stems, leaves and flowers do not lie dormant, but may germinate in water or in very humid atmospheres. Spores of parasitic fungi such as rusts germinate without the assistance of stimulants from their hosts. Spores of many facultative parasites germinate more vigorously in the presence of nutrients, such as sugars and amino acids. The sources of these nutrients may be physical wounds, decaying plant parts such as fallen petals and intact surfaces.

Spores are affected by substances which diffuse from or through intact cuticles. Drops of water incubated on plant surfaces for 1–2 days can be collected, combined and tested. Electrical conductivity will have risen indicating that electrolytes, presumably inorganic ions, have accumulated in the drops. Depending upon the plant part, the drops will have changed in their activities in spore germination tests. Drops held on rose petals become highly stimulatory to the germination of spores of *Botrytis cinerea*. Stimulants from petals are sugars, such as sucrose, glucose and fructose, to which *Botrytis* spp. are very responsive. This phenomenon may explain the ease with which these fungi attack flower heads in wet summers. Drops held on broad bean leaves often become partially inhibitory to germination. Rather little is known of the role of inhibitory substances which exude from plant surfaces.

In nature, many fungal spores land on a plant surface but very few of the spores seem to be able to parasitize the plant. The fate of these spores is largely unknown. Important questions are (i) do they germinate, (ii) if not, why not, (iii) do they lack nutrients or are they stopped by inhibitors, and (iv) if they germinate, can the germ-tubes grow? Answers to these questions can be sought in relatively simple experiments involving observation of spore behaviour on host and non-host leaf surfaces, and testing the effects of exudates on spore germination.

Before germ-tubes of parasitic fungi penetrate plant cuticles, they produce adhesive infection structures on the surfaces. Simplest infection structures or appressoria are slight swellings of the germ-tube, which may become coated with an adhesive mucilage. Complex structures are small mounds of branching hyphae over the infection point, but these are more typical of root-parasitic fungi. Fungal hyphae can be stimulated to produce very similar structures when they contact hard surfaces, such as glass. There is also evidence that chemical stimuli can affect the formation of appressoria.

Some very interesting observations have been made on the behaviour of parasitic fungi on leaves and stems during the time preceding penetration. For example, it is often observed that fungi which penetrate cuticles do so after forming appressoria over the junctions between cells. Hyphae of *Rhizoctonia solani* often grow over stems and hypocotyls of plants along the lines of junction between epidermal cells, as indicated in Fig. 5–2. Fungi, which enter through stomata, form appressoria over the walls

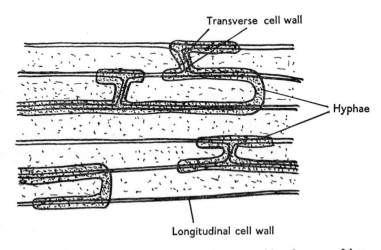

Fig. 5–2   Hyphae growing over cell walls on outside of stems of lettuce seedling.

of guard cells, and not elsewhere on the leaves. The nature of the stimuli which cause these reactions are not known. The physical nature of surfaces or the diffusion of stimulants may favour fungal activity in these places.

5.2   Entry into the host

Three routes of entry for parasitic fungi are through wounds, natural openings (such as stomata) and directly through intact surfaces. Different fungi have their own specialized ways of entering plants.

It is simple to show that sugars, amino acids and other substances exude in large quantities from wounds, and that these compounds stimulate growth of many parasitic fungi. However, spores of *Penicillium digitatum*, for example, cannot parasitize oranges even if supplied with nutrients for growth unless the fruit surface is physically wounded. Whether the wound circumvents a physical or chemical barrier is not certain. Wounds can be caused by climatic features such as hail, frost and wind, by animals and particularly by man. Wound parasites are very important in potato tubers and fruits such as apple and citrus.

Many obligate parasites enter plants through stomata. The problem of how these fungi detect stomatal apertures has been referred to above. Appressoria are formed over the openings. Narrow hyphae pass through and swell out to form sub-stomatal vesicles, from the which internal mycelium develops.

Direct penetration of cuticle is achieved by many facultative parasites and by powdery mildews, which push haustoria into epidermal cells from the surfaces of leaves and stems. Penetration follows adhesion of appressoria to the plant surface. The infection hyphae are very narrow compared with normal hyphae, and are difficult to detect except in histological studies of the highest quality. The fine infection hyphae usually emanate from the centres of under-sides of appressoria. Most infection hyphae pass through the cuticle and the walls of the epidermal cells. A few fungi such as *Venturia inaequalis* pass through the cuticle but then grow between the cuticle and the epidermal cells.

The appearance of appressoria and fine infection hyphae suggest that penetration is a physical process. The backward thrust, while the hypha is pushing through the cuticle, may be taken by the appressorium which is stuck to the cuticle. Positive evidence in favour of mechanical penetration is based on experiments with *Botrytis cinerea*. This fungus penetrated thin gold film and paraffin wax membranes. Penetration of wax membranes was prevented by increasing their hardness.

The possibility that penetration of cuticles can be an enzymatic process arises from claims that degradation of cuticle can be seen after some type of infection. Cuticle is a complex mixture of fats, waxes and other compounds such as phenols. There is evidence that the fat content of cuticles on leaves covered with mycelium of powdery mildews is much lower than that on healthy leaves. There is no reason why enzymes capable of degrading cuticle should not exist. In fact some saprophytic fungi which degrade leaf litter in soils produce enzymes which hydrolyse some parts of cuticle. Future research is likely to expand knowledge of enzymatic action on cuticles. Valuable supporting evidence may come from electron microscopic studies of penetration points.

Failure of fungi to parasitize some plants may reside in physical or chemical properties of cuticles. There is relatively little good positive evidence in favour of this. Some correlations have been suggested between increasing thickness of cuticle and increasing resistance to parasitic attack in some plants. On the other hand, the shrub *Euonymus* has an exceptionally thick cuticle and a high susceptibility to a powdery mildew. Some of the many chemicals in cuticles are antifungal, but they may not be present in sufficient concentration to prevent the localized types of penetration which occur. Failure of infection hyphae to develop within penetrated plants may be based on substances arising from affected cells beneath the cuticle and not on chemicals originating in the cuticle.

5.3   Growth inside the host

With the exception of powdery mildews which grow along outer surfaces of plants and push haustoria into epidermal cells, all parasitic fungi grow between or into host cells. A few parasites such as *Venturia inaequalis* grow between the cuticle and the epidermis.

Most obligate parasites grow between and push haustoria into parenchymatous cells. Electron micrographs of haustoria produced by powdery mildews, downy mildews and rusts show similar structures in all these types of fungi. The basic features of the haustoria and the invaded cells are shown in Plates 8 and 9. There is no evidence of any major derangement of host cells, walls or membranes. Enzyme action by the parasites, if involved in haustorial penetration, must be localized on the surfaces of the hyphae.

Facultative parasites are more destructive. They often grow into and between host cells, and cause damage in advance of their growth. These fungi produce extra-cellular enzymes which separate and degrade cell walls. Fungi such as *Botrytis cinerea*, *Sclerotinia fructigena* and parasitic species of *Penicillium* produce cell-separating enzymes in culture. The medium required is as follows:

1·0 g	pectic compound (sodium polypectate or pectin)
0·1 g	glucose
0·2 g	mycological peptone
0·1 g	KH_2PO_4
0·05 g	$MgSO_4 \cdot 7H_2O$

in 100 ml water adjusted to pH 6·0 with 0·1N NaOH or HCl.

This medium should be dispensed in 50 ml volumes in medical flats, autoclaved and inoculated with pieces of agar+mycelium. The growth period should be 5–7 days. The culture fluid is then collected, filtered and centrifuged, if possible, at 1000 g for 15 min. Samples of culture filtrate at pH 6·0 should cause softening of thin slices of potato tuber. Disintegration of tissue, when teased with dissecting needles, should occur within 1 h, but it may be necessary to incubate samples overnight, in the presence of a few drops of toluene which prevents bacterial growth.

Different parasites produce several different enzymes which act on different sites in the pectic and cellulosic parts of cell walls. Cell separating enzymes of a few parasites have been purified, and have been found to act exclusively on pectic substances. The material between cells and parts of primary cell walls are made of pectic substances.

A simple method of showing the activity of pectolytic enzymes is a cup-plate assay performed with the following medium:

1·0 g	sodium polypectate
0·05 g	ammonium oxalate
0·2M	mixture of KH_2PO_4/Na_2HPO_4 at pH 6·0
2·0 g	agar

in 100 ml water.

Twenty-millilitre volumes are poured into petri dishes. After setting, several 8 mm diameter holes are cut out with a sterile cork borer. The holes are filled with culture fluid and incubated for about 24 h at 37°C. The dishes are then flooded with 5N HCl. Clear haloes will appear within minutes around those holes which contained active enzyme, as in Fig. 5–3.

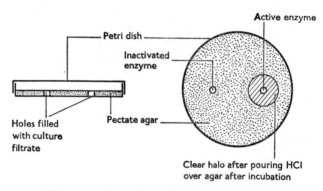

Fig. 5–3   Cup plate assay for pectolytic enzymes.

Parasitic fungi may secrete other enzymes which hydrolyze protein and fats in the host cells. Solubilized materials are absorbed by the hyphae as food supply.

Some facultative parasites produce non-enzymatic toxins which may play important roles in the suppression of host metabolism and ability to react against the parasite. Two of the most interesting toxins discovered to date are victorin and periconin. Victorin is a peptide produced by a species of *Helminthosporium* which can only parasitize a certain variety of oats. Victorin is toxic at a concentration of 1 part in 10 million to susceptible varieties of oats. Much higher concentrations of victorin are needed to affect resistant varieties of oats. This remarkable host specificity is also shown by periconin, a toxic peptide produced by *Periconia circinata*. Periconin is only toxic to those varieties of sorghum which are susceptible to attack by the fungus. It seems that victorin and periconin are the agents of specialized attack by these two parasites. Varieties of host plant resistant and susceptible to these parasites must differ in the possession of receptor sites for the specific toxic peptides.

In addition to tissue degrading enzymes and toxins, parasitic fungi may produce growth-regulatory substances (hormones) in their host plants. Some fungi secrete indole-acetic acid, the best known plant hormone, into culture media. In a number of plant diseases, some of the symptoms are similar to those caused by hormones. Some fungi cause marked distortions and over-growths in their hosts. See Plates 10, 11 and 12. Wilt fungi cause adventitious root formation and leaf epinasty. In the diseased plants, concentrations of indole-acetic acid have been found to be many times higher than in healthy plants. It is quite likely that much of the extra hormone is produced by the parasite, but ways are also known in which parasites can interfere with the processes controlling amounts of hormone in plants. Another plant hormone, gibberellin, was first found

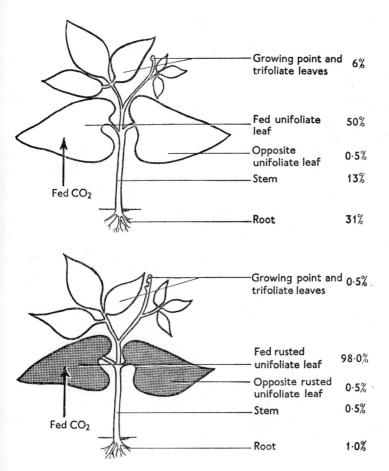

Fig. 5-4   Translocation of carbon-14 from healthy and rusted unifoliates of bean fed $^{14}CO_2$ in the light (source as Fig. 5–5).

in culture fluids of the parasite *Gibberella fujikuroi*. This fungus causes rice plants to become tall and spindly, and presumably does so by secreting gibberellins in the plant. Since this discovery in Japan some 35 years ago, gibberellins have been recognized as important natural growth-regulants in higher plants.

The advantage to the parasite of changed hormone concentration in the plant is not known. It may be concerned with the suppression of those features of host metabolism which attempt to resist parasitic attack. Alternatively the changes may be side effects of successful parasitism, and of no consequence to the parasite.

5.4 Reproduction of the parasite

The successful parasite completes its colonization of the host and produces units for its further propagation. The propagules may be sclerotia or resting spores which pass into the soil as the diseased root or stem decays. Leaf and stem parasites often have two types of propagule, a rapidly produced asexual spore used in spread of the parasite in the growing season, and a sexually produced spore which lies dormant throughout the winter.

The type of stimuli which cause parasites to switch their development from vegetative growth to formation of propagules are not clearly defined. Factors important in culture are light, temperature and exhaustion of nutrients in the medium. These factors are probably operative on parasites in their natural habits. An interesting example of this problem is the behaviour of black stem rust in wheat. Towards the end of the summer, pustules which have been forming uredospores switch to the formation of thick-walled black teleutospores. What governs the change? Possibly the fungus responds directly to physical climatic changes. Conceivably the fungus responds to major metabolic changes in the wheat associated with ripening of the grain and straw, and the onset of senescence.

Parasitic fungi probably draw heavily on their hosts for nutrients at the time of sporulation. Some studies on the fate of radioactive carbon dioxide photosynthesized by bean plants parasitized by rust are interesting in this respect. The radioactive carbon accumulated around rust pustules at the expense of growing points, roots and stems, the normal recipients in healthy plants. See Fig. 5–4 and 5–5. Furthermore the radioactivity passed predominantly to the fungal sugar trehalose and to fungal sugar alcohols and not to host sucrose. These fungal compounds are important reserves of carbohydrate in uredospores.

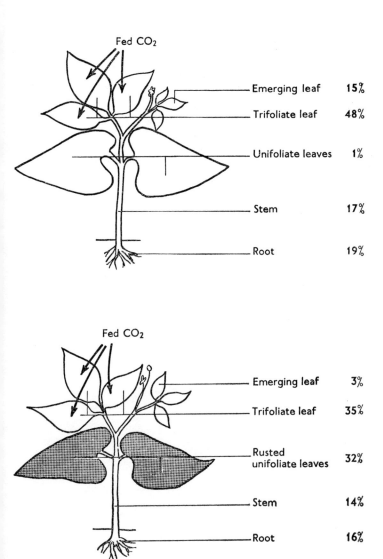

Fig. 5-5 Translocation of carbon-14 in healthy and rusted bean plants after feeding $^{14}CO_2$ to healthy trifoliates in the light (redrawn from LIVNE, A. and DALY, J. M., 1966, *Phytopathology*, **56**, 170–175).

44

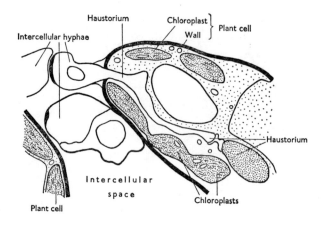

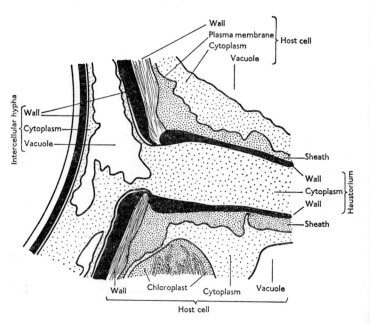

Plate 8 Electronmicrograph of haustorium of *Uromyces appendiculatus* in cells of bean leaf, *Phaseolus vulgaris* (from MR. N. V. HARDWICK, MR. A. D. GREENWOOD and PROF. R. K. S. WOOD, Imperial College).

Plate 9 Electronmicrograph of penetration point of haustorium of *Uromyces appendiculatus* through cell wall of bean (from MR. N. V. HARDWICK, MR. A. D. GREENWOOD and PROF. R. K. S. WOOD, Imperial College).

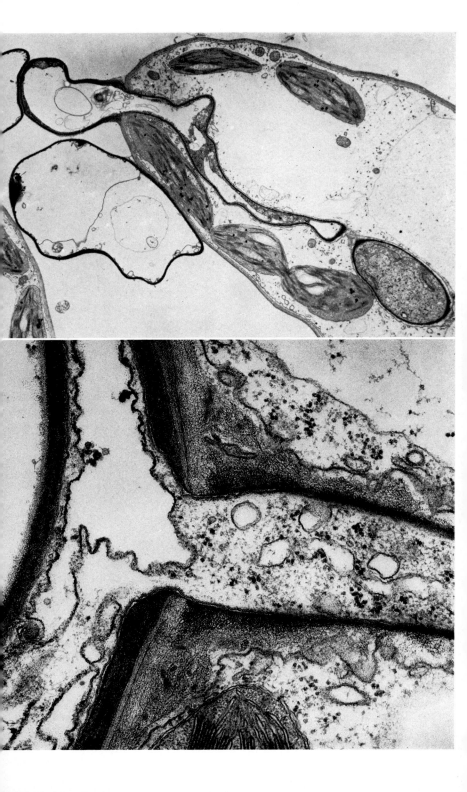

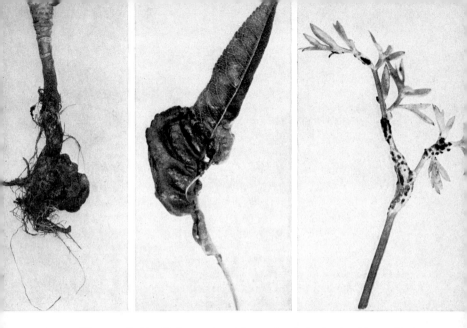

Plate 10 (*left*)   Club-root in cabbage after infection by *Plasmodiophora brassicae*.   **Plate 11** (*centre*)   Growth disturbances in leaf of *Prunus* caused by *Taphrina* sp.   **Plate 12** (*right*)   Swelling of leaves and petioles of earthnut (*Conopodium majus*) after infection by the rust, *Puccinia tumida*.

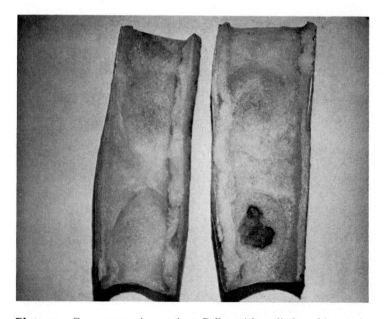

Plate 13   Cross protection against *Colletotrichum lindemuthiamum* in opened bean pods. Upper cavities were exposed to spores of a non-parasitic race, while lower cavities were treated with water. After 24 h the treatments were replaced by water in the left cavities and spores of a parasitic race in the right. The picture was taken 13 days later when typical symptoms had appeared in the lower right cavity but not in the upper cavity (from MR. R. A. SKIPP, Imperial College).

Resistance to Parasitic Attack 6

6.1 General Resistance

Spores of parasitic fungi are widely distributed, and many must come within the area of influence of many different plants. However each parasite is able to parasitize one or very few species of plant. Attempts usually fail to make a fungus colonize plants other than their particular hosts. Therefore it is clear that most plants have means of resisting attack by most parasitic fungi.

Presumably this observed specialization of host/parasite relationships has resulted from natural selection of genetic recombinants and mutants among populations of plants and fungi. Thus, for example, *Albugo candida* is restricted in ability to parasitize members of the Cruciferae, and each rust fungus is restricted to a particular species of host plant.

6.2 Varietal resistance

Man has selected particular varieties of crop plants or garden flowers for intensive cultivation. These varieties have been chosen for many different properties including resistance to specialized parasites. Gardeners may know that new varieties of *Antirrhinum* are available with resistance to antirrhinum rust. Commercial growers are very familiar with this type of change in most crop plants.

Plant breeders produce resistant varieties in two ways. One method is to select and breed on from those individual plants which do not succumb to attack during epidemics of particular diseases. The other method is to seek related species or varieties of crop plants which are resistant to parasites in greenhouse or field trials. Resistance can be bred from the resistant plants into the crop varieties so that resistance is combined with desirable qualities such as yield and appeal to the buyer. Sources of resistance to particular parasites have often been found among wild plants in parts of the world where the crop plant originated. Some useful sources of resistance to *Phytophthora infestans* in potato have been found in Central and Southern America, the original home of the potato.

In the course of breeding programmes, it is often necessary to understand the genetic basis of resistance. This is done by crossing appropriate resistant and susceptible plants, and collecting the resulting seed. These seeds are then sown, and the resulting plants are allowed to self-pollinate or are cross-pollinated. From the proportions of plants which are resistant or susceptible to the parasite, deductions are made about the genetic control of resistance. Simple Mendelian ratios are often observed, and it is common to find that resistance to a parasite is conferred by a single gene. However there are many cases of polygenic resistance, which require much more extensive genetic analyses.

The process of evolution of specialized parasites is continuous, and new forms of parasites often arise and overcome the resistance which man has bred into his crop plants. This process has been very troublesome in the case of varieties of cereals resistant to rust fungi. Common experience is that a few years after a variety resistant to prevalent forms of the rust has been introduced, a new form of the rust evolves to parasitize the resistant variety. The problem is so complex that several hundred forms or *physiologic races* of black stem rust are now known. These races are defined by testing their ability to parasitize standard *differential varieties* of wheat. Evolution of physiologic races has occurred in many different species of parasitic fungus. An example of differentiation of races of a parasite is shown in Fig. 6–1.

6.3   Gene-for-gene relationships

Some studies of the genetics of host resistance have been accompanied by studies of the genetics of parasitic specialization. Precise genetic relationships have been found between some rust and powdery mildew fungi and their host plants. Different single genes in varieties of flax confer resistance to different physiologic races of flax rust. Single genes in the races of rust confer ability to parasitize host varieties. Thus, changes in single genes in the rust seem to have enabled the parasite to overcome the appropriate single genes bred into the flax varieties as sources of resistance.

Genetic control of resistance and parasitic ability is more complex in many host/parasite relationships. Man may be wise to seek polygenic control of resistance for his new varieties. Monogenic resistance may be overcome too readily by chance mutation or genetic recombination in parasites. Polygenic resistance may delay the evolution of new physiologic races of parasites.

6.4   Processes of resistance

Apart from the fact that resistance is controlled genetically, so little is known of the processes of resistance that generalizations are difficult to make. However one general problem will be considered first. There may be a process of resistance to parasitic fungi which is common to all plants, but there may also be some mechanisms which are peculiar to certain plants and parasites. The large number of scientific papers on resistance contain a variety of claims about mechanisms of resistance. However most of these papers concern simple forms of resistance, which have been easily studied but may be unusual and peculiar. Examples, such as the resistance of coloured onion bulbs to attack by most parasitic fungi, are described below. The types of process which underlie physiologic race/variety specializations of parasites for hosts have been little studied. It is possible that advances in this more difficult area of research will reveal a process of resistance common to most parasite/host interactions. With conjectures of

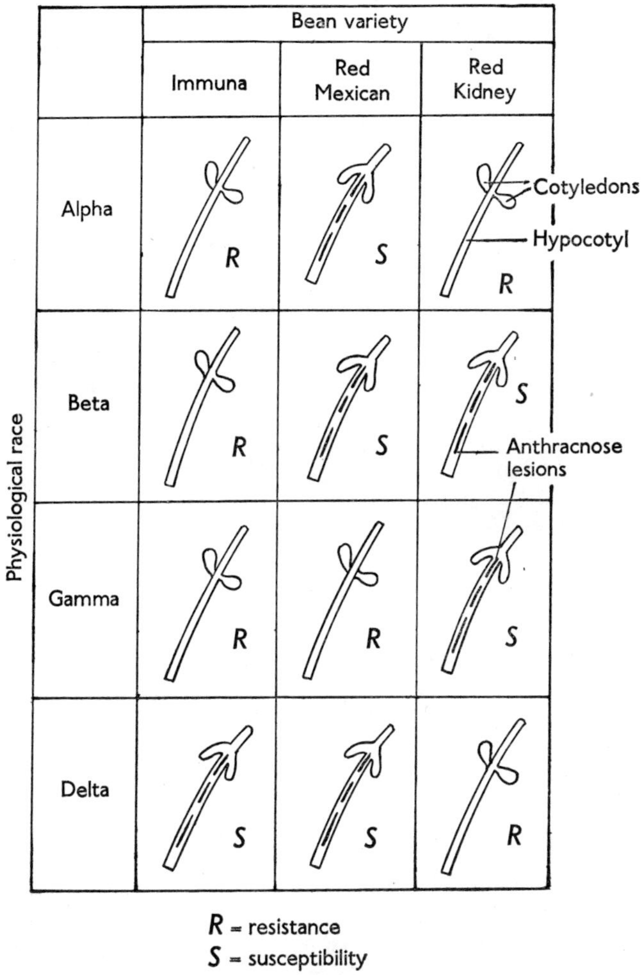

R = resistance
S = susceptibility

Fig. 6–1 Physiologic races of *Colletotrichum lindemuthianum* differentiated by their interaction with varieties of bean. Bean seedlings were sprayed with spore suspensions of the parasite and incubated in humid chambers for ten days, when the indicated symptoms were observed on the hypocotyls depending upon race and variety.

this type in mind, the following discussion has been divided to consider some proven mechanisms which may be peculiar, and to show the type of work which may reveal general processes of resistance.

6.5 Peculiar mechanisms of resistance?

One of the most thoroughly studied mechanisms is that whereby coloured onion bulbs are resistant to most of the fungi which normally parasitize onion bulbs. These fungi readily colonize exposed inner fleshy parts of the coloured bulbs. The resistance lies in the dead outer scales Drops of water incubated on these scales accumulate potent inhibitors of spore germination and germ-tube growth of *Colletotrichum circinans* and some other parasites of onion. The inhibitors are the phenolic compounds, catechol and protocatechuic acid shown in Fig. 6–2. They diffuse from the dead scales where they exist in high concentration. These phenols are not coloured, but merely happen to be associated with the pigments. Breeding resistance into other onion varieties results in the transfer of the phenols to the recipient varieties. A fungus, *Aspergillus niger*, which can parasitize coloured onions is not sensitive to the phenols. This is a clear explanation of the general resistance of a plant, but is based on the content of unusual tissues, dead scales.

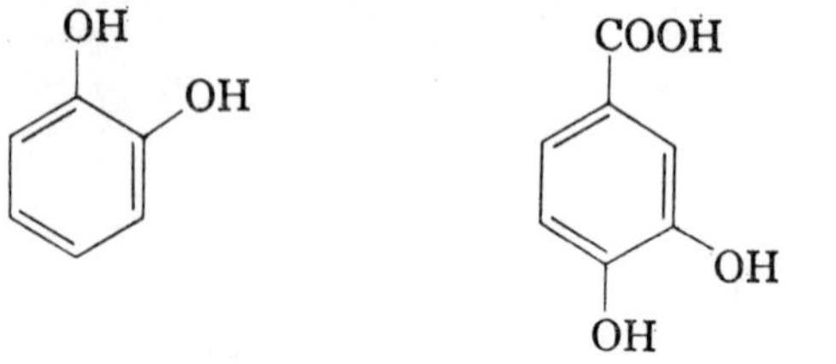

Catechol	Protocatechuic acid

Fig. 6–2   Structures of antifungal compounds in outer scales of coloured onion bulbs.

Many suggestions have been made that substances, often phenols, present in healthy cuticles or cells confer resistance to particular parasites. One of the best examples concerns the failure of *Ophiobolus graminis* to parasitize the roots of oat, although it readily attacks the roots of other cereals. An aqueous extract of tips of oat roots inhibits the growth of the parasite. The inhibitory substance is a glycoside called avenacin, which has a complex chemical structure. A variant of the parasite, *O. graminis* var. *avenae*, is able to parasitize oat roots and produces an enzyme, avenacinase, which hydrolyses avenacin to much less toxic compounds.

It has often been suggested that a plant may be resistant if it fails to supply a nutrient requirement of the parasite. Positive evidence of this form of resistance to naturally occurring fungi, as distinct from induced mutants, is restricted to one case, that of parasitism of the bark of plum trees by *Rhodosticta quercina*. The fungus is deficient for the vitamin, inositol. The bark content of inositol is high in a susceptible variety of tree, but low in a resistant variety.

It seems most unlikely that resistance to facultative parasites can often be based upon absence of an essential nutrient. By definition, these parasites grow on simple nutrients, which abound in living plant tissues. All resistance mechanisms based upon such features as thick cuticles, presence of antifungal compounds in healthy plants and absence of essential fungal nutrients are simple. These mechanisms seem to be unable to underlie the highly specific reactions between many varieties of a plant and many races of a parasite. Philosophically it is difficult to conceive that a plant can possess, for example, a multitude of antifungal compounds each specifically effective against one race of a parasite.

6.6 General processes of resistance?

In addition to the arguments outlined above, there are several basic reasons why general processes of resistance should be sought in plants. One reason is that plants seem to respond actively to infection by many parasites, and not passively as would occur if plants relied on such mechanisms as in section 6.5. A second reason is that the resistant condition is not usually dependent upon a property of the plant alone, but upon an interaction of host and parasite based on their genetic contents.

Active responses of resistant plant cells to some fungi can be seen. After contact or penetration by germ-tubes of the fungus, the host cells and often their neighbours react in a violent way, known as an hypersensitive response. Disorganization, browning and apparent death of these cells occur, as in Fig. 6–3. This type of response can be seen in epidermal and upper palisade cells of broad bean leaves infected with *Botrytis cinerea* as described in section 2.3. As the hypersensitive reaction takes place, the fungus becomes limited to the point of penetration. Mere death of host cells seems unlikely to cause the limitation of growth of facultative parasites. Some process accompanying the hypersensitive response must cause the limitation. Many hypersensitive reactions occur when obligate parasites such as rusts and powdery mildews infect uncongenial hosts. The reactions may be invisible to the unaided eye, but through a microscope they can be seen to be limited to very few cells. Hypersensitive death of host cells around hyphae of obligate parasites may be sufficient to cause resistance.

Active responses of plants to infection include increases in respiratory rate and changes in many metabolic reactions. Increased rates of respiration can be easily studied in plants infected with powdery mildews collected in garden or field during summer months. Alternatively these

changes can be detected in broad bean leaves with 1-day old lesions caused
by spores of *Botrytis* as described in section 2.3. Discs from diseased
leaves should be cut from parts of leaves covered with mildew mycelium
or from areas bearing brown lesions caused by *Botrytis*. Discs must not
be allowed to dry out before adding them to reaction vessels. Normal
techniques of Warburg manometry at 20–25°C are used with approximately

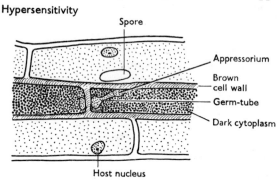

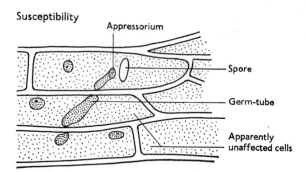

Fig. 6–3 Hypersensitivity and susceptibility of cells of bean hypocotyls to
races of *Colletotrichum lindemuthianum*, observed 3 days after spraying with
spores (redrawn from LEACH, J. G., 1923, University of Minnesota Agricultural
Experiment Station, Technical Bulletin 14).

0·2 g fresh weight of discs of 5 mm diameter cut with a cork borer. Discs
should be added to 2 or 3 ml of 0·1M KH_2PO_4 per flask, before 0·2 ml
of 10% KOH and a wick of a folded 1·5 cm square of filter paper are placed
in the centre well. To prevent photosynthesis flasks must be shielded
from light in the Warburg bath, for example by wrapping them in alumin-
ium foil. Dry weights of discs in each flask should be found after the
experiment. Results are best expressed in terms of Q_{O_2}, i.e. μl O_2/h/mg
dry weight of tissue. Values of 1–2 are normally obtained for healthy leaves,
and values for diseased leaves of the same ages are several-fold higher. A

further refinement of the experiment can be made with powdery mildews. These fungi are superficial and can be removed from the leaves with a small brush. The respiratory rates of brushed, healthy and mildewed leaves can be compared. It will be found that, although all mildew except for haustoria in epidermal cells has been removed, the diseased leaves respire almost as rapidly as those with mildew remaining. This suggests that much of the increased respiration occurs in the host cells. More difficult experiments involving the peeling-off of epidermis from healthy and diseased leaves support this view.

Similar types of study have been made with leaf discs to measure the rate at which radioactive metabolites, such as glucose, carbon dioxide and plant hormones, are used. Rates of metabolism of all these compounds are much affected by the presence of parasitic fungi.

These experiments merely show that the host responds actively to infection. That active responses of similar sorts occur, when resistant cells are infected, has been shown in many experiments with less readily available plants. This type of research has not advanced far enough to show how these active responses are involved in the process of resistance.

Further support for the view that resistance is an active process comes from experiments in which host activities are suppressed temporarily by anaesthetics or heat treatments. *Sclerotinia fructicola* does not normally parasitize bean tissue. When bean tissue was inoculated with this fungus after exposure to 44°C for 2 h or to ether vapour for 24 h, it was colonized. The host tissue was not killed by the treatments, and it recovered its resistance after an interval of 72 h between treatment and inoculation.

A number of experiments suggest that a process of cross-protection can be achieved with plants and parasitic fungi, like a limited form of acquired immunity in animals. One series of experiments was done with potato tubers and isolates of *Phytophthora infestans*. The results suggested that attempted infection by isolates which were non-parasitic on a certain variety protected those tubers from infection by normally parasitic isolates. The protection was non-specific being effective against quite unrelated parasites of potato. The protection was confined to the precise areas of the tuber inoculated with the non-parasitic isolates. Similar demonstrations of localized cross-protection have been made with other plants and parasites, such as with bean and races of *Colletotrichum lindemuthianum* as in Plate 13.

These are the main reasons why active processes of resistance to parasites in plants should be sought. Now it is necessary to consider the little that is known about these processes, and to suggest the areas of research which may be most productive in solving these problems.

6.6.1 Phytoalexins

The term phytoalexin was proposed to mean a warding-off substance produced by a plant. A phytoalexin can be defined as a substance produced by a plant after infection and responsible for preventing further growth by the infecting fungus. The existence of phytoalexins was suggested to account for the cross-protection reaction between potato tubers and *Phytophthora infestans* described above. However no substance was isolated from infected tubers as a phytoalexin. The first demonstration of a phytoalexin as an extractable substance was in work with bean (*Phaseolus*) pods infected by *Sclerotinia fructicola*. Pods were opened and seeds removed to expose the large expanse of pod tissue, which is not covered by thick cuticle nor likely to have been exposed to previous attack by microorganisms. When spore suspensions were added to seed cavities, the spores were observed to germinate readily and to cause visible hypersensitive reactions within 24 h in the host cells. Droplets of spore suspension were collected and combined after different time intervals. The volumes of liquid were made cell-free by centrifugation and then tested for their action on fungal growth in culture. It was found that an antifungal compound began to accumulate in droplets in seed cavities 14 h after spore suspensions were pipetted out. The droplets became progressively more antifungal with time. The fungitoxicity could be extracted from combined droplets with petroleum spirit, and seemed to be caused by a single chemical compound.

The compound was not identified until after further similar work with pea pods and the same fungus. The compound from pea was isolated, identified and called pisatin. The compound from bean was found to be closely related chemically and was called phaseollin. The structures are shown in Fig. 6–4.

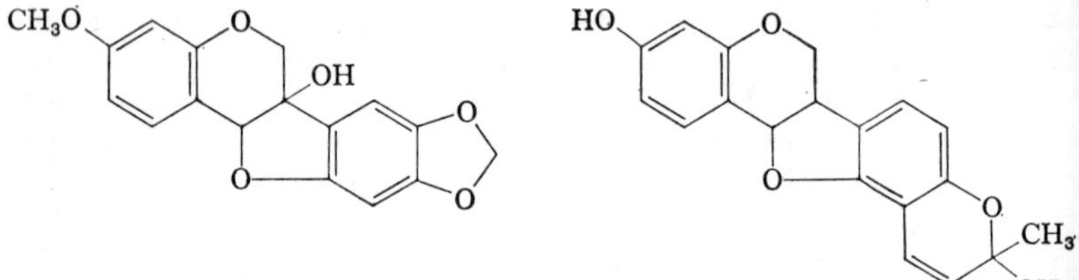

Pisatin   Phaseollin

Fig. 6–4   Structures of pisatin and phaseollin.

Because of ease of experimentation, peas and pisatin have been extensively studied. Fungistatic amounts of pisatin are formed by pea tissues in response to many fungi, parasites and saprophytes, to culture filtrates of fungi and to many chemicals such as copper ions. Very little pisatin is found in healthy cells and in physically damaged cells.

While these discoveries were being made, other research groups had found that plant parts such as carrot roots and orchid and sweet potato tubers produce antifungal compounds after infection by fungi. Some of the compounds are broadly similar to pisatin, others are quite different. All seem to be formed by the plants in response to adverse treatments, as well as to infection.

The phytoalexins in the plants studied may be responsible for the general resistance of those plants to most fungi. The role of phytoalexins in controlling the development of specialized parasites is less certain, and has been examined in only two host/parasite relationships. One of these concerns *Rhizoctonia solani* which causes small lesions on bean stems. These lesions are green and water-soaked at first and, within a few days, they become dark-brown but limited to their original size. Two phytoalexins, phaseollin and an unknown phenol, accumulate in the green lesions in amounts which prevent fungal growth. There is little doubt that these phytoalexins limit the parasite to the area of the initial lesion. The second relationship is that between *Botrytis* and broad bean leaves, as described in section 2.3. When the fungus infects, the leaf cells produce a phytoalexin which accumulates around the penetration area. *Botrytis cinerea* is sensitive to the phytoalexin and is prevented from growing further in the leaf. *Botrytis fabae* is less sensitive and removes the phytoalexin, and then grows slowly through the leaf. Therefore phytoalexins seem to be very important in controlling the growth of parasites in these simple relationships with legumes.

Little is known of the role of phytoalexins in physiologic race specialization. Many people find it difficult to accept a role in these situations for such readily formed compounds which have little specificity in their effect on fungi. Known phytoalexins are produced in response to many types of biological and chemical damage. An important finding would be that some parasitic fungi had specific ability to prevent or suppress phytoalexin formation in their host. Unequivocal evidence of this is not yet available. Such a finding would encourage further study of the roles of phytoalexins in other plants and in some of the most specialized host/parasite relationships.

6.6.2 *Protein reactions*

The high specificity of some host/parasite relations might be based upon reactions between proteins in the two partners. Very little has been discovered in this aspect of biology, but two distinct possibilities are clear.

The restriction of physiologic races of parasites to particular host varieties is similar to the compatibility of certain pollen grains and styles, and to the acceptance of tissue grafts only by recipient animals closely related to donors. A common theme to investigations on processes underlying such specificity is that surface proteins on contacting partners may be important in determining what follows. Experiments with flax rust compared proteins in races of the rust and in varieties of flax. Results suggested that races of rust must possess some proteins common to their host varieties if they are to be successful in colonizing these varieties. Possibly some form of information matching occurs when fungal and host cells contact. This information might be in surface proteins. The host cell may not react against the parasite if the fungal surface has similar proteins to the host surface. If the proteins on surfaces fail to match, then the host cell may react violently. In fact the reaction may be the observed hypersensitive response, and this alone may be enough to stop growth of the rust, an obligate parasite. This is highly speculative and many experimental studies are needed to investigate the possibility.

Another way in which proteins of host and parasite may interact is through an antibody type of reaction. In animals, many carbohydrates, lipids and especially proteins act as antigens when injected into the body, and cause the production of proteinous antibodies. These antibodies are so constructed that they specifically counter the particular inducing antigen. A second injection with a particular antigen causes a very much more rapid production of the antibody. This reaction is the basis of acquired immunity to infectious agents by either previous exposure or immunizing injections. It has often been suggested that plant cells may be able to do something similar, perhaps locally in an infected cell and its neighbours. In fact the first search for phytoalexins may have been guided by a desire to find antibodies. The known phytoalexins are not proteins and lack the specificity of animal antibodies towards their inducing agents. Of course plants may not be able to make specific antibodies, and phytoalexins may be generally analogous to animal antibodies. However the possibility that antibodies are formed in plants needs testing, and past failure may have happened because it is difficult to get active proteins from plants. Experience in extracting active mitochondria, viruses and enzymes from plants shows that these entities are easily inactivated by tannins. Tannins are the coloured materials which form on cut surfaces of apples or potato tubers on exposure to air. Tannins are common in infected cells. Necessary precautions should be taken in searches for active antibodies in plants, and until adequate studies are done final conclusions cannot be drawn.

6.6.3   Nucleic acid metabolism

Obviously high specificity between hosts and parasites is based on the genes, and therefore on the nucleic acids of both partners in the relationship. The interaction between host and parasite which can prevent growth of the parasite may occur at any stage of metabolism. The interesting possibility that this interaction may occur at the level of nucleic acid metabolism has been proposed, at least for races of obligate parasites.

Obligate parasites, by definition, have a deficiency of some form which can only be supplied by their living host cells. Conceivably the deficiency is for a component of nucleic acid metabolism and/or protein synthesis. The required material might be soluble RNA. If the host cell supplied the component, this might explain the obligate nature of these parasites.

If a particular race required a specific soluble RNA from a specific host, this might explain specificity as well as obligatism.

This attractive idea requires supporting evidence, but in fact it has received the opposite in the last two years. First, the best-known obligate parasite, *Puccinia graminis tritici*, has been grown on artificial medium, although there are doubts about a component of the medium, the vigour of the resulting spores and the races of rust which behave in this way. If this rust proves to be a facultative parasite, despite many past failures of attempted culture, then it has no special deficiency. Secondly spores and germ-tubes of the rust seem to be able to synthesize protein without any external contribution of substances such as soluble RNA. Progress in both these areas of research is so recent that they may need re-assessment in a few years' time. They do not support the idea of a specific requirement from their hosts by obligate parasites.

6.6.4 A two-phase process of resistance

Whatever general process of host/parasite specialization becomes proven, it seems valid to suggest a two-phase process, except in the event of deficiencies of soluble RNA as in section 6.6.3 or simple relationships as in section 6.5.

The first phase of contact of fungal and host cells may be a recognition reaction, involving the matching of information. This is probably mediated by nucleic acids or proteins. Usually the recognition reaction will fail, and a rejection reaction will follow. Rejection may be mediated by hypersensitive death of host cells or imperfect association of haustorial and cytoplasmic surfaces in the case of obligate parasites. Rejection may be caused by phytoalexins or cellular antibodies. On few occasions, the recognition reaction will permit the parasite to colonize the host and reproduce itself. The capacity to avoid provoking the host or to specifically suppress the hosts' resistance mechanism may be an important feature of a parasite.

Further Reading

AINSWORTH, G. C. and AUSTWICK, P. K. C. (1959). *Fungal Diseases of Animals.* Commonwealth Agricultural Bureaux, Farnham Royal, Bucks.

AINSWORTH, G. C. (1961). *Ainsworth & Bisby's Dictionary of the Fungi.* Commonwealth Mycological Institute, Kew, Surrey.

AINSWORTH, G. C. and SUSSMAN, A. S. (ed.) (1965–68). *The fungi. An Advanced Treatise. Vol I. The Fungal Cell. Vol II. The Fungal Organism. Vol III. The Fungal Population.* Academic Press, New York & London.

ALEXOPOULOS, C. J. (1962). *Introductory Mycology.* John Wiley & Sons Ltd., New York & London.

AMERICAN PHYTOPATHOLOGICAL SOCIETY (1967). *Sourcebook of Laboratory Exercises in Plant Pathology.* W. H. Freeman & Co., San Francisco & London.

BURNETT, J. H. (1968). *Fundamentals of Mycology.* Edward Arnold, London.

CARTWRIGHT, K. ST. G. and FINDLAY, W. P. K. (1958). *Decay of Timber and its Prevention.* Her Majesty's Stationery Office, London.

COMMONWEALTH MYCOLOGICAL INSTITUTE (1968). *Plant Pathologist's Pocketbook.* Commonwealth Mycological Institute, Kew, Surrey.

DADE, H. A. and GUNNELL, J. (1966). *Class Work with Fungi.* Commonwealth Mycological Institute, Kew, Surrey.

GARRETT, S. D. (1956). *Biology of Root-infecting Fungi.* University Press, Cambridge.

GREGORY, P. H. (1961). *The Microbiology of the Atmosphere.* Leonard Hill (Books) Ltd., London & Interscience Publishers, Inc., New York.

HOLTON, C. S. et al. (ed.) (1959). *Plant Pathology Problems and Progress, 1908–1958.* University Press, Wisconsin.

HORSFALL, J. G. and DIMOND, A. E. (ed.) (1959–60). *Plant Pathology, an Advanced Treatise. Vol I. The Diseased Plant. Vol II. The Pathogen. Vol III. The Diseased Population, Epidemics and Control.* Academic Press, New York & London.

STAKMAN, E. C. and HARRAR, J. G. (1957). *Principles of Plant Pathology.* The Ronald Press Co., New York.

WALKER, J. C. (1969). *Plant Pathology.* McGraw-Hill Book Co., Inc., New York, Toronto & London.

WILSON, M. and HENDERSON, D. M. (1966). *British Rust Fungi.* University Press, Cambridge.

WOOD, R. K. S. (1967). *Physiological Plant Pathology.* Blackwell Scientific Publications, Oxford and Edinburgh.